Dialogues and Games of Logic

Volume 7

Negotiating with a Logical-Linguistic Protocol in a Dialogical Framework

Volume 1
How to Play Dialogues: An Introduction to Dialogical Logic
Juan Redmond and Matthieu Fontaine

Volume 2
Dialogues as a Dynamic Framework for Logic
Helge Rückert

Volume 3
Logic of Knowledge. Theory and Applictions
Cristina Barés Gómez, Sébastien Magnier, and Francisco J. Salguero, eds.

Volume 4
Hintikka's Take on Realism and the Constructivist Challenge
Radmila Jovanović

Volume 5
The Dialogue between Sciences, Philosophy and Engineering. New Historical and Epistemological Insights. Homage to Gottfried W. Leibniz 1646-1716
Raffaele Pisano, Michel Fichant, Paolo Bussotti and Agamenon R. E. Oliveira, eds. With a foreword by Eberhard Knobloch

Volume 6
Meaning and Intentionality. A Dialogical Approach
Mohammad Shafiei

Volume 7
Negotiating with a Logical-Linguistic Protocol in a Dialogical Framework
Maria Dolors Martínez Cazalla

Dialogues and Games of Logic Series Editors
Shahid Rahman   shahid.rahman@univ-lille3.fr
Nicolas Clerbout
Matthieu Fontaine

Negotiating with a Logical-Linguistic Protocol in a Dialogical Framework

Maria Dolors Martínez Cazalla

Research project sponsored by Leffe Abbey

© Individual author and College Publications 2019
All rights reserved.

ISBN 978-1-84890-321-0

College Publications
Scientific Director: Dov Gabbay
Managing Director: Jane Spurr
Department of Informatics
King's College London, Strand, London WC2R 2LS, UK

www.collegepublications.co.uk

Original cover design by Laraine Welch

All rights reserved. No part of this publication may be reproduced, stored in a retrieval system or transmitted in any form, or by any means, electronic, mechanical, photocopying, recording or otherwise without prior permission, in writing, from the publisher.

To my family

The following book is an edition of *College Publications* and the *Axe Transversal Argumentation* headed by Shahid Rahman at the University of Lille 3–Charles-de-Gaulle and in collaboration with the laboratory *Savoirs, Textes, Langage* (Unité Mixte de Recherche 8163, Lille 3).

Negotiating with a Logical-Linguistic Protocol in a Dialogical Framework is the result of years of reflection. Some time ago, while working in commodities, the author felt how difficult it was to decide the order in which to use arguments during a negotiation process. What would happen if we translated the arguments into cards and played them according to the rules of the Bridge game? The results were impressive. There was potential for improvement in the negotiation process. The investigation went deeper, exploring players, cards, deals and the information concealed in the players' announcements, in the cards and in the deals. This new angle brought the research to Neuro-Linguistic Patterns and cryptic languages, such as *Russian Cards*.

In the following pages, the author shares her discovery of *a new application for Logical Dialogues: Negotiations*, tackled from basic linguistic structures placed under a dialogue form as a cognitive system which 'understands' natural language, with the aim to solve conflicts and even to serve peace.

Table of contents

Foreword xi
Acknowledgements xv
INTRODUCTION 1
ACT I: The Bridge Game 5
    Chapter 1.1: Bridge. Is it a useful game to make rational decisions in the world of negotiations? 7
        The Bridge Ontology 7
        The Bridge Epistemology 9
        Does Bridge belong to the *Game Theory*? 11
        Is Bridge a useful game to make rational decisions in the world of negotiations? 12
    Chapter 1.2: Bridge application to *Camp David Accords* 13
    Chapter 1.3: The Bridge. A bridge toward Negotiations. Lights and Shadows 23
ACT II: Neuro-Linguistic Patterns (NLP) 27
    Chapter 2.1: Are NLP a useful tool to make rational decisions in the world of negotiations? 28
    Chapter 2.2: How do Linguistic Patterns work in our mind? 29
    Chapter 2.3: Shadows under the light of NLP 32
ACT III: Cryptic Language: *Russian Cards* 35
    Chapter 3.1: Hunting Hackers. Is it useful to think like a 'mute guest' to prepare a negotiation talk thoroughly? 35
        A Dialogical Semantics for the *Russian Cards* 36
          1. Mathematical rules for the *Russian Cards* 36
          2. Particle rules 37
          3. Structural rules for a game played in the intuitionistic Logic framework 38
          4. Formalisation for the general case 40
          5. Interpretation keys 41
          6. Practical cases 42
    Chapter 3.2: Two lessons from the 'mute guest' 49
    Chapter 3.3: A Dialogical Framework. A light in the darkness 51
CONCLUSION: Negotiating with a Logical–Linguistic Protocol
                      in a Dialogical Framework 55
Bibliography 61
ANNEX: Study Case: *Camp David Accords* 69
Author 109

Foreword

Maria Dolors Martínez-Cazalla is a pioneer. She is a pioneer in negotiation, and she shares three traits with the pioneers in navigation: boldness, creativity and passion.

Boldness. Before focusing on basic linguistic structures, Maria Dolors Martínez-Cazalla was a business woman. Her experience in this field was certainly crucial to realize that presenting arguments in a specific order matters. Scholars and practitioners involved in negotiation are well aware of the crucial importance of the arguments emphasized by the parties. However, they are now asked to go one step further in considering not only the content of the arguments, but also their particular sequences. To Maria Dolors Martínez-Cazalla, the vital question is: in which order should the arguments be used and under which syntactic structure should be expressed in order to make a difference? To address this question, she translated arguments into cards and played them according to the rules of Bridge. Seeing the potential of this initial step, she pursued her objective with determination and started exploring linguistics and logic in order to test various scenarios. Her path is undoubtedly original. In choosing it, Maria Dolors Martínez-Cazalla took risks. She can now enter the room of daring scholars.

Creativity. *Negotiating with a Logical-Linguistic Protocol in a Dialogical Framework* is one of the most innovative books devoted to the negotiation process. Many writers have explored the underlying factors behind the success or failure of negotiations. As early as 1716, François de Callières was studying negotiations. His book *De la manière de négocier avec les souverains* follows in the tradition of Machiavelli's *The Prince* in the advice it gives readers. But rather than reducing negotiations to preparing for war, he describes them also as a harbinger of peace. Since then, many handbooks have followed in succession. These works stress in particular the importance of preparation and the formal aspects of negotiation. Questions of status, choice of a particular language, setting calendars, mandates, and agendas often condition the results of a process. Most of these books present negotiating techniques, tactics, and strategies, referring sometimes to standards of rhetoric, argumentation, and persuasive processes. However, none of them underlined the significance of basic linguistic structures as conjunctions, disjunctions, or conditionals.

In this regard, no one could deny that Maria thought about something unthought. The use of the cards metaphor is not radically new. Negotiators and mediators are often compared to players, while their moves, their interactions and their deals are described in terms of games. Nonetheless, only a few experts took the metaphor seriously. Game theorists mobilize economic theory (for studying comparative costs), and social psychology (analyzing cooperative and competitive behavior) for this purpose. They insist on the actors' rationality, and ask fundamental questions about their choices, the conditions affecting those choices, and the level of trust between parties. This strategic approach is often associated with the study of the prisoner's dilemma that led to the development of a theory of cooperation based on a form of "conditional trust" (tit for tat). From this perspective, the most compelling long-term strategy is to start by cooperating and in the next phase to respond in the same way as the other party.[1] Maria Dolors Martínez-Cazalla's theory is also based on rationality and calculations. Nonetheless, her research question is completely original.

Expert in Logic and specialist in the rules of Bridge, she enlightens three of the most important phases of any negotiation process: the preparation phase, the information phase and the argumentation phase. The order of the arguments can actually be taken into account as soon as the information needed to make a diagnosis is gathered (preparation phase). Once the diagnosis has been made, the parties generally enter the second major stage designed to develop a jointly agreeable formula that will serve as a referent for an agreement. During the information phase, long discussions then strive to determine the terms of the exchange. The third phase deals with the details of the transaction. This argumentation phase is precisely the moment for fine-tuning positions, calibrating concessions, and specifying the terms of the exchange leading to the agreement's finalization. The care taken in each of these phases determines to a great extent whether or not the agreement reached will be implemented. At each of these phases, the logical-linguistic protocol emphasized in this book invites parties to pay attention to the best order in which they can use their arguments and to build them under the most favorable syntactic structure for them.

Passion. Since our first encounter in 2010, Maria Dolors Martínez-Cazalla never stopped working to push boundaries. Far from any kind of repetition, her theory raises new questions. And, even though we cannot predict the

---

[1] See Robert Axelrod, *The Complexity of Cooperation*, Princeton (NJ), Princeton University Press, 1997.

progress of negotiations through her protocol alone, they complete the major tools that are available to fully grasp the complexity of negotiation processes. There is much at stake. Historians and ethnologists agree that all human societies are characterized by negotiation. Certain realities might seem to be non-negotiable a priori. Beliefs, values, and identities are rarely the result of compromise. They are by nature non-divisible and unlikely to be modified by any dealings. Similarly, notions of justice and truth do not seem open to bargaining, at least in principle. It is therefore not rare for parties to immediately affirm the non-negotiable nature of certain positions in the framework of negotiations that are predominantly conflictual rather than cooperative.

Even if the theory of Maria Dolors Martínez-Cazalla is conceived for all kinds of negotiation, and not only for diplomatic negotiations, let me illustrate my point by a concrete example which has been burning for years now on the international scene, namely the peace talks on Syria. In this specific case, the removal of Syrian President Bachar Al-Assad was long qualified as non-negotiable by representatives of the Syrian opposition. However, the length of the conflict, the intervention of foreign powers such as Russia and Iran, and leadership changes in third parties (whether in the United States or France) seem to have shifted the inviolability of this red line. In the Middle East, issues regarding the right of return and holy places have also prompted positions presented as non-negotiable. Yet the deadlocks created by these problems do not mean there is no imaginable solution. Value conflicts certainly prove more intense and harder to settle than interest conflicts. But one should never rule out a priori that experienced negotiators and/or mediators might succeed in turning value conflicts (religious or identity-based ones, for instance) into interest conflicts. That being the case, it seems problematic to present certain subjects as inherently non-negotiable. Certain realities —although presented and perceived as such for decades— may over time be subject to transactions, depending on the different actors' circumstances and objectives. The work of Maria Dolors Martínez-Cazalla shows us that these variables (circumstances and objectives) are not the only ones that deserve to be considered.

Boldness, creativity and passion. These traits allowed Maria Dolors Martínez-Cazalla to contribute to a better understanding of the subtle art of negotiation. To some, negotiation is all about power. To others, it is all about relationships. A focus on a dialogical framework can help the parties to find themselves on the same side of the fence. The exercise might seem risky and discouraging —but it is always compelling. Although it can be dizzying, it is no doubt worth remembering that, in the end, "it is all about reaching a compromise. A matter of imagination"[2].

Dr. Valérie Rosoux
Senior Research Fellow FNRS
UCLouvain
valerie.rosoux@uclouvain.be

---

[2] Francis Walder, *Saint-Germain ou la négociation*, Paris, Gallimard, 1992.

Acknowledgements

They try to kill off competition;
Then, bankrupt take an expedition,
For, having slashed their prices downward,
Their best direction's out of townward;
If one of them won't cut the price
His brothers won't be overnice,
For if the customer is tight,
The goods he gets will serve him right;
The masters now make shoddy wares:
If they can sell'em then who cares?
But such a business will not keep:
To by stuff dear and sell things cheap;
Some tradesmen underestimate
A deal, then take the city gate.
Whoever loves bargain, he
Won't get much for a guarantee.
They makes so much a little cost;
They sell it quick and nothing's lost;
They profit from a slick veneer
Now honest tradesmen disappear.[3]

I would like to thank all the people who have taught me the game of negotiation, even when the *honest tradesmen disappear*. However, they are just a possible (as an option) element of the game. What is really important and exciting is the game itself. I thank my family and my teachers for teaching me how to play, to love the game of Life: the art of trading and negotiating with Life every moment of my life. Thank you family and teachers for teaching me to enjoy Life.

I would particularly like to say a special word of gratitude to Professor Dr Ángel Nepomuceno at the University of Seville and Professor Dr Shahid Rahman at the University of Lille-3, and also to Professors Dr Valérie

---

[3] S. Brant (1494), *The Ship of Fools* (quoted by Marsh, 1984, p. 387).

Rosoux and Dr Pierre Dehez at the Catholic University of Louvain; all of them have made this work possible, without them I could not have done it. Thank you for your trust, support and teachings. Above all, thank you for making me a better person.

I would like to say thanks to Mrs Lola Nieto, my master in the art and science of Bridge. Thank you for your science converted into art.

I am grateful to Ms Juani Romera, who tuned me up in the specific mathematics I needed for this work. Thank you for your devoted time.

I would especially like to express a word of gratitude to Ms María Cantarero for being a sort of personal assistant to me. Without her dedication and care I could not have made it. Thank you for walking, even running, up and down for me.

Very big thanks to my sponsor, Leffe Abbey, for making my last stretch easy and sweet.

I would also like to thank my students: Alvi, Andrés, Javi, José Domingo, Marcos, Sebastián and Toño. Their questions improved my reflection. Thank you for challenging me.

I would like to thank all my friends too, for their support and for putting up with me during this time. Thank you for being my relief.

My most sincere thanks to Dr Asteria Albert and Fr José-Ignacio García, SJ. Thank you for the revision of the English language.

Finally, a grateful recognition to Ms Laura Hidalgo. Thank you for your help with the layout; it made the difference.

> Even if
> *What has been, that will be; what has been done, that will be done.*
> *Nothing is new under the sun!*
> Ecclesiastes 1:9[4]
>
> *There may be other knowledges to acquire, other questions to consider,*
> *starting, not from that what others have known,*
> *but from what they have ignored.*
> S. Moscovici[5]

---

[4] United States Conference of Catholic Bishops (2011).
[5] Our translation for: "Puede que actualmente haya otros conocimientos que adquirir, otras cuestiones que plantearse, partiendo, no de lo que los demás han conocido, sino de lo que han ignorado" S. Moscovici (quoted by Morin, 1986, p. 20).

INTRODUCTION

Even if "nothing is new under the sun", "there may be other knowledges to acquire, other questions to consider, starting, *not* from that what others have known, *but* from what they have ignored [emphasis added]", by choice or by chance –that is not relevant; the matter is to think about something un-thought to this moment. This is the hazardous passion developed here, and you are invited to share it.

Negotiating with a Logical-Linguistic Protocol in a Dialogical Framework is the result of years of reflection. Some time ago, while working in commodities, we felt how difficult it was to decide the order of arguments[6] (*arguments* understood as explained in the footnote) used during a negotiation process. As in a Bridge game, we translated the arguments into cards and played them according to the rules of Bridge, and saw how it worked to deal with them as in a Bridge hand. The results were impressive. We were thrilled about the potential for improvement in the negotiation process. We decided to investigate deeper in the possibility to undertake negotiations applying Bridge rules to organise the order of arguments. This was the subject of a previous work (2011): *The Bridge. A bridge toward Negotiations*.[7] That work was the first formal attempt to establish a protocol to know the best order in which to use arguments during a negotiation process by converting them into cards and playing a Bridge hand. However, as it will be shown later, the study revealed some limitations. This subsequent work is an attempt to reduce these limitations.

The new work had to be more scientific and precise, so the decision was taken to start a research in Logic as the best framework and tool to address this subject. Following this path, the idea of turning arguments into cards to play a Bridge hand expanded progressively and went one step ahead exploring players, cards, deals and the information concealed in the player's announcements and in the cards and/or in the deals. This new angle brought the research to Neuro-Linguistic Patterns –NLP– and cryptic languages, like *Russian Cards* developed by van Ditmarsch (Van Ditmarsch, van der Hoek, & Kooi, 2008, pp. 97-104 et 108).

---

[6] We take always the word *argument* along this work as the conclusion of an inferential process. Thus, *Argument* = Premise/s + Reasoning = *Conclusion*.
[7] Martínez-Cazalla (2011), Master's thesis available at the Library of Economic, Social, Political Sciences and Communication (BSPO), at the Catholic University of Louvain (UCL).

Finally, this work is an attempt to think how to create logical dialogues to tackle negotiations, meaning: solving conflicts from basic linguistic structures (conjunctions, disjunctions, conditionals) placed under a dialogue form as a cognitive system which 'understands' natural language and where there is a permanent feedback between both.

This piece of research aims to show and to share just a path, not a conquered territory, toward negotiating in a dialogical framework, and remains always open to any possible refinement. It has been developed like a tragedy in three acts. Each act has been a 'conceptual mimesis' of the arguments used during the negotiation process to produce a 'catharsis', an improvement in the negotiation process. The three acts have a spiral form, the first one is Bridge, the second is Neuro-Linguistic Patterns and the third is the cryptic language *Russian Cards*. The procedure for this research has been to study each part in the sight of its contribution and its limitations. Our task has been to address our focus to the limitations, with the aim to reduce them at every step. The Conclusion shows only a possible map, a guide to choosing the order of arguments and their syntactic structure, as will be expressed/expounded in negotiations.

The structure of each act, as a step in this path, will be:
1. A presentation on the appropriateness and accommodation of the specific subject in the whole research.
2. Application to prepare negotiations.
3. Lights and shadows, or some interesting considerations to keep in mind for planning negotiations.

In order to judge our success in establishing a protocol to know the best order in which to use arguments during a negotiation process, we needed a testing ground in which to verify our conclusion. This is why we chose a completed negotiation case to guarantee an objective application, because there is no possibility to change the events The case chosen for study as a model for this 'experiment' is *Camp David Accords*. An exhaustive analysis of this case is found in the Annex,[8] so that the testing ground is available to assess this research in terms of right or wrong, because without this document the semantic truth would be unknown. To preserve the rigour and the aseptic nature of this research, we have *not* applied any framework susceptible of being applied later —we will discover its real usefulness throughout the research. Thus, you will not find any application of the *Game Theory*, neither NLP nor Dialogical Semantics along this analysis of *Camp David Accords*. A

---

[8] A published presentation by Martínez-Cazalla (2012) about this subject is also available.

preliminary research had been carried out on this negotiation, as mentioned above. The same study case has been chosen with the purpose to give continuity to the already started investigation. This was the reason for selecting this particular case, and *not* the idea that this theory is only applicable to cases of international negotiations. This theory is for negotiations, whatever they are, regardless of their scope.

The methodology used here has been the one that is appropriate in Logic: many paper reflections, thought drafts –not included here because they do not have a decisive character for the final result–; specific sources[9] such as manuals, books, articles and documents about the different subjects tackled along this research; and personal reflection comparing the different results and information. Working with specific sources, we tried to remain as rigorous as possible while opening a new theoretical path into negotiation analysis and into the field of applied Logic (one of the difficulties of opening a new path, no matter how fascinating it may be, is that no sources exist while it is being built). As you know, the instrumental nature of Logic was recognised as early as Aristotle's *Organon*. In fact, Logic has been a tool for philosophical studies since Aristotle, although many logicians see Logic only as a family of formal systems. Logic is *not* applied to philosophical problems in the same way an engineer may apply some techniques. Nevertheless, many logical notions transcend the particular formal systems and Logic can offer there a rigorous language –with precise meanings– to study philosophical discourses and discourses in social and human sciences. Besides, it is a great help for enhancing precision in communications. As we see, modern Logic deals with a wide range of intelligent interactions across academic disciplines, from humanities to natural sciences. This dynamic turn involves the logical Dynamics (dialogues as a form of reasoning, dialogical Logic, Study of Knowledge, communication process, etc.). In this sense, van Benthem was clever when saying:

> *Logical Dynamics is a way of doing Logic* [emphasis added], but is also a general stance. Making actions, events, and procedures first-class citizens enriches the ways in which logic interacts with philosophy, and it provides a fresh look at many traditional themes. Looking at logic and philosophy over the last century, key topics like quantification, knowledge, or conditionality have had a natural evolution that went back and forth across disciplines.

---

[9] Note that in this work there are two sections named *Bibliography*: one at the end of the general part and another attached to the Annex. There are specific references in each; other references are shared sources for both research approaches, and therefore appear repeated in both lists.

They often moved from philosophy to linguistics, mathematics, computer science, or economics—. (Van Benthem, 2011, p. 268)

Thus, there is a multitude of fields in which Logic can be applied. Thank you for your interest in this path toward using applied Logic on a new field: *Negotiations*.

This work consists of a creative and an innovative effort full of risks. The 'experiment' will confirm whether that innovation and risk were worthy and the reader will judge on the degree of accommodation.

In the following pages you will discover *a new opportunity to apply Logical Dialogues, this time to deal with Negotiations*, to solve conflicts (as objective application), and even to serve peace (as a subjective option, since tools do not have an ethical value in themselves).

ACT I: The Bridge Game

As has been said before, this work is the continuation of a previous one, *The Bridge. A bridge toward Negotiations* (Martínez-Cazalla, 2011). Thus, the next pages will be a sort of summary of the above, in order to enable a right support for the present reflection. We will use its most relevant data for our present research. Anyhow, the complete text is available to be consulted at the Library of Economic, Social, Political Sciences and Communication (BSPO) at the Catholic University of Louvain (UCL).

In our previous research about Bridge and negotiations, we conceived negotiations as a double spiral, like DNA, and Bridge as its linkage. One of the spirals is the theoretical negotiation axis that supports a possible rational explanation about the implementation of negotiations; we named it the *scaffold*. Consequently, the natural limit of the scaffold was its implementation, which we named the *scaffolding*, the other spiral, which supports the negotiation itself –once the theory is implemented, it is *not* possible to change the events. Both spirals are simultaneous and intertwined; they cannot be understood separately and they exist always together, they are the two sides of the same coin.

The scaffold was analysed in the theoretical-deductive way: we started from theory to achieve a practical application. All negotiations start with a decision, the decision of negotiating about something; Chapter 1 was an analysis of the *Decision Theory*. Given that this analysis was necessary but not sufficient because we needed the decision to be rational, we showed that the most rational decision would be the intersection point between the '*f(optimization)*' and '*f(satisfaction)*'. As satisfaction cannot be objectively calculated (the level of satisfaction is always subjective), the rationality of the decision remains on the side of the objectivity, that is, on optimization. Both functions are mixed in games: the result of a game speaks about its optimization and the players speak about their satisfaction. Therefore, our second step was the analysis of *rationality* by means of *Game Theory* –Chapter 2. Once we completed the initial requisites of a negotiation, which include the decision of negotiating something and its rational basis, we were ready to analyse the negotiation from a theoretical approach with the purpose to answer the implicit question in this chapter: *Is the Game Theory useful to make rational decisions in negotiations?* –Chapter 3. Up to now we have been building the foundations, the first three steps of the theoretical negotiation spiral. We concluded that the rationality of decision-making in the course of a negotiation can be addressed by the rational study of a game that represents the negotiation. *What game in particular will be studied?*, and *Why are we choosing this game and not another?* This was the content of our next chapter, where we dealt with the fourth and last step of our scaffold –Chapter 4. Then it was the time to answer the question: *Is Bridge a useful game to make rational decisions in the world of negotiations?*

From this scaffold, we recover Chapter 4 turned here into *Chapter 1.1: Bridge. Is it a useful game to make rational decisions in the world of negotiations?* The aim of this chapter is just to know and understand the rationality that is implicit in Bridge –and not to become expert Bridge players. Therefore, the length and depth of this chapter is restricted accordingly.

For the other spiral, the scaffolding, the procedure toward knowledge was empirical-inductive, starting from a specific case and reaching a feasible theory. Therefore, we tackled this practical spiral from a study case: *Camp David Accords*. This long, dense, comprehensive chapter –Chapter 5– was analysed in the most sceptical way. The goal was to know whether Bridge was a possible internal bridge –linkage– between the theoretical framework –scaffold– and the practical cases –scaffoldings. In order to accomplish that, we needed to study a past case (future cases cannot possibly prove a hypothesis) and analyse it following an approach *not* based on the *Game Theory*. We did so because we suspected that Bridge would be the linkage. Otherwise, the degree of accommodation between the two spirals could have been conditioned.

You will find former Chapter 5, the study case (scaffolding) *Camp David Accords*, as *Annex*.

Once we had the two spirals defined, we could ask about the possible linking element. The element proposed as bridge was the Bridge game. Thus, the next part was an application of the Bridge game to a case of negotiation. We chose to apply Bridge to *Camp David Accords*, making it possible to verify whether Bridge was a feasible bridge to the analysis of negotiations. In other words, we wanted to confirm that Bridge could be a possible bridge between the two spirals. Therefore, this section –Chapter 6– was an application of Chapter 4 (present Chapter 1.1) on Chapter 5 (present Annex). Our implicit questions here were: *Is the Bridge game the element that bridges both spirals in negotiations? Could the Bridge game be a tool for a rational analysis of negotiations?*

We retrieve here the study of the spirals linkage [–former Chapter 6–] as *Chapter 1.2: Bridge application to Camp David Accords*.

Finally, the Conclusion was a reflection about the following: *To what extent is it possible to consider Bridge as an element for a rational analysis of negotiations? What light does it provide to our understanding of negotiations? Which are its limitations?*, and *How could we overcome these limitations?*

We bring up these reflections here, turned into *Chapter 1.3: The Bridge. A bridge toward Negotiations. Lights and Shadows*. This chapter will *not* be a duplicate of the former Conclusion, since this second look at the subject has added some more light, as will be clarified through the pages of this book. At the end of this chapter, we will be able to suggest what the greatest shadow is, to which we aim to throw some light along this research work.

Act I, composed by three chapters, will be the path to discover why the Bridge game is useful to make rational decisions in the world of negotiations and where its limitations are.

Chapter 1.1: Bridge. Is it a useful game to make rational decisions in the world of negotiations?

[This chapter is taken from Martínez-Cazalla, 2011, pp. 28-35. You will find in square brackets the additions and/or alterations to the original text].

We suggest studying Bridge [...] because we [think] [...] it is [the most appropriate game] to make rational decisions in the world of negotiations. *Why do we [think so]?* Because, since the *Game Theory* is based on the analysis of the [games as microcosms of situations of true conflict], and Bridge is a [...] game [of communication, strategy, and battle] (cf. Kast, 1993, p. 11), [...] [Bridge] could [...] be the best reflection of a negotiation. For the purpose of answering this question it will [be] necessary to analyse Bridge, following a step by step process, so we can prove Bridge is a useful game to make rational decisions and, more specifically, it is [a clear] illustration [of] what happens in a negotiation. [In order] to do this, we will answer three questions: *What is its ontology? What is its epistemology?* and [...] *Can we say that Bridge belongs to the Game Theory?*

The Bridge Ontology

Before starting with the analysis of the Bridge ontology we need [to offer a definition] of *ontology*. Ontology is "(...) a branch of metaphysics concerned with the nature and relations of being".[10] In the case of the Bridge game, *Which is the nature and relations of its being?* To answer this question we will use a [more] modern definition of *ontology*: "(…) an explicit specification of a conceptualization" (Gruber, 1993, p. 1). Therefore, we need to describe the underlying *conceptualization* in the explicit Bridge *specification*.

The explicit Bridge specification responds to its conceptualization. This is the relation by means of a deck of cards equally shared among all players. The nature of [the] relations [among] [...] [the] players appears covered [under] its conceptualization, the conceptualization of a common language [shared by the] players. Therefore, the ontology [...] we are talking about is a *semantic ontology*, since "A semantic ontology is a conceptualization, common to a community of agents that understand natural language, of the categories and relations that pervade the agents' environment as a whole. It can be used to specify the logical form as the truth-functional meaning of agent messages embedded in natural language" (Schneider & Cunningham, 2003, p. 1403). [Thus], we [should] analyse the Bridge language to know its ontology, its being.

The Bridge language is common to its four players, but the players are divided into two pairs (named: North-South, the host pair, and East-West, the visiting pair). The relation between pairs is non-cooperative, but [within] the pair they keep a cooperative relation. Through a conceptualized language (common to the community of players) the players reach an accord about which will be the trumps (no-trumps is [also a] possibility of trumps). The [object of the game] is to reach the best communication possible between players of the same pair that use a common language, which is also known by the other pair. [All announcements must be true, lies are not allowed]. The objective is to reach the maximum of possible tricks from the probabilistic and combinatorial calculus. Apparently, Bridge is a non-cooperative game [...], but in Bridge the winner is the pair that has reached the 100% of the possible tricks [in more hands]; for that, [a kind of collaboration between the pairs is strictly necessary, therefore the meta-Bridge game is a sort of *win-win* negotiation]. [In an ideal hand, which is quite common in Bridge], when a pair [proponent] obtains a [surplus trick, it] is *not* because it has played very well, it is always because the other pair [opponent] has played wrongly; and when a pair [proponent] does *not* obtain all

---

[10] Definition of [the] word *ontology* in: *Encyclopædia Britannica*, 1st meaning. Retrieved from http://www.britannica.com/bps/dictionary?query=ontology [We are choosing the definition from a generic source, as the *Encyclopædia Britannica*, and *not* from a particular theoretical framework, in order to preserve the rigour and the aseptic nature of this research].

the possible tricks [,it] is [...] [because it has played wrong, and *not*] because the opponent has played very well. Therefore, in Bridge there is always an intrinsic collaboration process. Every obtained trick [over] or [below] the average will be a 'gift' (over) or a 'loss' (below), but [that is *not* the aim of the game, except for some cases where tricks may be won or lost as a result of the probabilistic skills used to play a particular hand combination. In these cases, the result over or below the expected will depend on the quality of the player's play. Such a scenario is not the usual; most Bridge hands are regular –ideal– ones, and these are the hands we are analysing here]. To understand this apparent paradox [–cooperative + non-cooperative game = *win-win* meta-game–] it is necessary to analyse the Bridge epistemology [first].

The Bridge Epistemology

We will start by offering a definition of *epistemology*. Epistemology is "(...) the study or a theory of the nature and grounds of knowledge especially with reference to its limits and validity".[11] If the *nature* of Bridge is founded on [...] mathematical principles, then in order to know the structure of Bridge it will be necessary to apply reasoning in accordance with [the principles of mathematical Logic, since] *Logic* is "(...) a science that deals with the principles and criteria of validity of inference and demonstration: the science of the formal principles of reasoning".[12] Therefore, our mission will be to discover the logic that lies behind its ontology, as "Formally, an ontology is the statement of a logical theory" (Gruber, 1993, p. 2). [Thus], our question shall be connected with the underlying Bridge logic, [which is] founded on the [mathematical principles of combinatorics and probability].

[...] Bridge is played by four players, and as the deck has 52 cards, [...] each player holds 13 cards. [These] 13 cards are obtained by random distribution ($C^{13}_{52} * C^{13}_{39} * C^{13}_{26} * C^{13}_{13} = 52! / 13! \, 13! \, 13! \, 13!$ –cf. Borel & Chéron, 2009, p. 38). This is the first step of the game and the only hazardous [one]. From here [on], all the other steps will be an application, more or less correct [–the degree of correction depends only on the player's knowledge and

---

[11] Definition of [the] word *epistemology* in: *Encyclopædia Britannica,* 2nd meaning. Retrieved from http://www.britannica.com/bps/dictionary?query=epistemology [We are choosing the definition from a generic source, as the *Encyclopædia Britannica*, and *not* from a particular theoretical framework, in order to preserve the rigour and the aseptic nature of this research].

[12] Definition of [the] word *logic* in: *Encyclopædia Britannica,* 1st meaning a (1). Retrieved from http://www.britannica.com/bps/dictionary?query=logic [We are choosing the definition from a generic source, as the *Encyclopædia Britannica*, and *not* from a particular theoretical framework, in order to preserve the rigour and the aseptic nature of this research].

qualities–], of [the mathematical principles of combinatorics and probability]. As the game implies [getting] the highest number of possible tricks, every time a player plays a card the others must play a card too. The played cards will be only those [that] fulfil the rules of the game: they must always [be] [...] of the same suit. [In case the players] do not have any, then there are two possibilities: either they play trump or they play [another] suit. [...] [The routine will be]:

[0.] Whoever starts the game will choose a card (C^1_{13}), and then the next players will have two possibilities:
1. If they have cards in the played suit, then they will choose from those in the played suit ($C^1_{n=\text{number of cards in the played suit}}$).
2. If they do *not* have cards in the played suit, then they will choose from [the] other cards. Again, there are two more possibilities:
    2.1. If [they] play at trumps:
        2.1.1. Then [they will have to] play a trump ($C^1_{n=\text{number of cards in trumps}}$).
        2.1.2. Then [they will have to] play a card of [another] suit ($C^1_{n=\text{number of cards in another suit}}$).
    2.2. If [they] play at no-trumps: [they will always have] the possibility to play any card (C^1_{13}).

The winner of the trick is the one who has played the highest [card] of the played suit, or who has played the highest [trumps card], if they [play at] trumps. If they are playing at no-trumps, then the winner trick will always be the highest [card] of the played suit. [...]

Evidently, [as the game goes on, there is] a smaller [choice for every trick] (C^1_{13-n} [n=] number of played [tricks], being 'n' in accordance with the chosen case (case 1 or case 2 –2.1.1.; 2.1.2. or 2.2). Therefore, in each trick we have less [choices for] the possible cases. [...]. This makes [it] really important to choose the card which is in 'perfect symmetry', [...] we need to know when to play [it] –once a card has been played, [this] card will *not* be available for the next [tricks]. This is the explanation [to] our paradox [–cooperative + non-cooperative game = *win-win* meta-game]. This necessity of 'perfect symmetry' is the key to obtain [...] 100% of the possible tricks, and [therefore it] is strictly necessary that each player plays the 'symmetric' card. We can say that Bridge is a cooperative game [...] hidden behind a non-cooperative face [both under the 'umbrella' of a *win-win* negotiation at the meta-game level].

Being Bridge a game with such complexity (non-cooperative between the pairs, but at the same time cooperative [within] each pair), *Can we say that Bridge belongs to the Game Theory?*

[Does] Bridge belong to the *Game Theory*

According to the analysis [of the] [...] *Game Theory* [(cf. Martínez-Cazalla, 2011, pp. 20-24)], we can conclude that Bridge belongs to the *Game Theory*. [This is a] *non-cooperative/cooperative* [game in which] *two collective parties* (two pairs) [deal with] *one issue* (the game itself). The goal is to achieve a specific effect by applying [a] strategy; this achievement [is based] on the probabilistic combinatorial mathematical calculus. [...] Hazard is *only* present at the moment [when the cards are distributed, and] *never* during the development of the game.

In fact, Bridge belongs to the category of the *ideal games* (cf. McClennen, 1992, pp. 47-60), [which characteristics are]:

(1) *Common knowledge*[13] [footnote added]. There is full common knowledge of (a) the rationality of both players (whatever that turns out to mean), and (b) the strategy structure of the game for all players, and the preferences that each has with respect to outcomes.

The force of this condition is that a player *i* knows something that is relevant to a rational resolution of *i*'s decision problem, then any other player *j* knows that player *i* has that knowledge. This is typically taken to imply (among other things) that one player cannot have a conclusive reason, to which no other player has access, for choosing in a certain manner. That is, there are not hidden arguments for playing one way as opposed to another.

In addition, one invariably finds that the analysis proceeds by appeal to the following (at least partial) characterization of rational behaviour for the individual participant.

(2) *Utility maximization*. Each player's preference ordering over the abstractly conceived space of outcomes and probability distributions over the events that condition such outcomes can be represented by a utility function, unique up to positive affine transformations, that satisfies the expected-utility principle.

(3) *Consequentialism*. Choice among available strategies is strictly a function of the preferences the agent has with respect to the outcomes (or disjunctive set of outcomes) associated with each strategy.

Following Hammond (1988), condition (3) can be taken to imply that strategies are nothing more than neutral access routes to outcomes (or disjunctions of outcomes); the latter are what preferentially count for the agent. In particular, then, if two strategies yield exactly the same probabili-

---

[13] [It should be noted that the term *common knowledge*, as understood here, does *not* necessarily have the same meaning it has in the context of epistemic Logic].

ties of the same outcomes occurring, then the agent will be indifferent between those strategies. (McClennen, 1992, pp. 47-48)

We can recognise the Bridge ontology and epistemology in these three characteristics of the *ideal games*. Therefore, we can assume Bridge [belongs] to the *Game Theory*. [...] Then, *Will it be a useful game to make rational decisions in the world of negotiations?*

Is Bridge a useful game to make rational decisions in the world of negotiations?

Following [...] [the] *Decision Theory* [(cf. Martínez-Cazalla, 2011, pp. 15-19)], the more rational decision would be found in the intersection point between the '*f(optimization)*' and '*f(satisfaction)*'. As the satisfaction cannot be objectively calculated (the level of satisfaction is always subjective) the rationality of the decision remains on the side of the objectivity, that is in the optimization, and, as [both functions are mixed in the games, and as the result of a game tells about its optimization], then the *Game Theory* [is] useful to find the intersection point. Furthermore, [according to] [...] [the] *Negotiation Theory* [(cf. Martínez-Cazalla, 2011, pp. 25-27)], if the rationality of the decision-making in the course of a negotiation can be addressed by the rational study of the game that represents the negotiation, then we can state that Bridge is a useful game to make rational decisions in the world of negotiations.

At this point it seems that our question is solved; but incisive minds [still] have another question: *Why have we chosen Bridge as [the] game to make rational decisions in the world of negotiations, and not any other game?* There are two answers, one [is] more theoretical and the other one [is] more connected with life.

- *From a theoretical point of view*, Bridge fulfils the characteristics [of] a game to make rational decisions in the world of negotiations. It is a cooperative game with two parties and [one] issue, [and] at the same time Bridge is a non-cooperative game. In short, although the ideal game to negotiate is a cooperative game (collaboration-compromise), [the bargaining is always a competition, anyway]. In fact, in almost all negotiations, the real situation is a *bargaining mix* (cf. Lewicki, Barry & Saunders, 2004, p. 38): [one part of the agreement is the consequence of collaboration and another part is the consequence of competition]. In addition, Bridge is a [non-random] game. It [shows] a perfect mathematical symmetry, where the use of 'dialogue' intra-pair and inter-pairs is a ruled system where there is a constant maintenance of common knowledge [–*common knowledge* as understood by McClennen (quoted in page 11), and *not* necessarily having the same meaning as in the context of epistemic Logic–] (cf. Kiel & Schader, 1994, pp. 171-180). Also, the fact that the winner is the pair that [has reached 100% of the possible tricks in more hands] entails that [players are very close to] the intersection point of optimization and satisfaction. [This means a *win-win* result at the meta-game level, because this score is the consequence of a strictly mathematical application of rules: nothing has been stolen from the opponent].

- *From [an experiential] point of view*, Bridge [is similar to life]. If we analyse each instant of our life, we [will] see that we always have a partner, [sometimes] even [a partner who remains] hidden, [...] that provides us with something [...] we have not, but [...] we [...] need in that precise moment of our life. At the same time, there is always somebody, [sometimes] even [a person who remains] hidden, who does not want [us to] we reach our goal. The elements that we have in favour or against our goal, [are the 'life cards'; depending on the way we play them, we] will achieve a higher, or lesser, degree of optimization and satisfaction in our own life. If we are strong in one of the elements, we will struggle to lean on it, and consequently we will [propose to] play at trumps; if we [have similar strength] in all the elements, we will try to lean on all, and then we shall play at no-trumps.

[...] We have chosen Bridge because [...] it [seems to be a very appropriate model] to make rational decisions in the world of negotiations and in the sphere of life. The game of Life is the art of trading and negotiating with Life at every moment of [our] life, and Bridge [could be] the best theoretical representation of the game of Life. [We will test it in the next chapter].

Chapter 1.2: Bridge application to [...] *Camp David Accords*

[This chapter is taken from Martínez-Cazalla, 2011, pp. 71-81. You will find in square brackets the additions and/or alterations to the original text].

[Before starting this chapter, you are asked to create a scenario for the application]:

Let us imagine that we are on August 1978. It is summer time, we are enjoying our holidays and the telephone rings. The prestigious editor in chief of a foreign affairs magazine –who knows our great love [for] Bridge and our great interest [...] [in] negotiations– proposes us the challenge of writing an article for [their] magazine. This article must be a prediction about what will happen, what will be the possible rational agreement at the [...] [Camp David summit] (scheduled [for] the next month, September 1978). [...] This article [has] a condition: the prediction can only be based on a virtual Bridge game played [by] the leaders [meeting] at the summit (Begin, Sadat and Carter). And we [do accept], perhaps because of the heat's drowsiness, or perhaps because we are crazy [for] logical challenges. (Martínez-Cazalla, 2011, p. 70)

[The objective information about what happened can be found in the Annex document. This information will always be available to corroborate the Bridge application to our study case].

The starting point [of] this chapter is two questions: *Is [the] Bridge game the element that bridges both spirals in [...] negotiations? Could [the Bridge game] be a tool for a rational analysis of [...] negotiations?* To answer [these questions], we will apply a hypothetical [Bridge game to] the 'future' possible agreement [in Camp David], [as we proposed at the introductory pages to this act]. Thus, the game starts and [the] only information available is the literature [on] the history [of] [...] relationships [among the parties] and their preparation for the meeting. Both, information and literature are very weak because at this time (*this imaginary game took place in August 1978*) the Israeli Government did not allow [public access to the statistical yearbook abstract, which was kept] strictly confidential. This is the reason [why] our data game cannot be founded on official sources. Therefore, this game is based on our personal background on this matter.

A Bridge game starts with the identification of the two pairs: the North–South pair (host pair) and the East–West [pair] (the visiting pair). *Who is North–South, the host pair?* The [pair which] plays at home, in this case, [is] Carter and Begin. Carter is the host because the negotiations will take place [in] the USA, and the USA is the historical ally of Israel. Then, *who is the East–West, the visiting pair?* The visiting pair is Carter and Sadat. The alliance between USA and Egypt is really weak: [from] a historical perspective Egypt is a country under the USSR influence –we are still at the *Cold War era*–, and from the perspective of [...] interests, although they are not [very many], [...] the USA needs guaranteed [...] the Middle East raw materials [while] Egypt needs the financial aid [of] the USA.

Both pairs have now been identified. [As we see], one of the players will play in both teams. *Is that possible?* In our case yes, because [one of the players –Carter for USA– is not interested in the object of negotiation] (the object is the condition to achieve [the] goal; in our case, [the object is] the occupied territories [from] the '67 War). [The USA only interest is to reach peace in the Middle East, the aim of this negotiation. Thus, their proper role is to be the mediator]. […] Then, as the USA do not take part directly on the object (necessary condition as mediator) and [as it is playing in both teams] (condition of sufficiency [in the case the mediator is not only a mediator, but he also has a direct interest in the goal achievement]), we can [omit] Carter from the game [proper]. [Thus, we are lucky because our game will be even simpler. Since the game deals with the object itself, and not with the objective, the goal is the consequence of the game but not the game itself]. Carter will be present in the game as a hidden force, reinforcing [the strength of each one of the other players but *not* directly playing].

Summarizing, we are [at the table and we] have two real players: Begin versus Sadat; we have 52 cards (a deck of cards) which represent the object that is at stake –the occupied territories during the '67 War–, [and] we have the goal: the peace in the Middle East.

[This] is the [time] to look at the cards. *Who has what?* Before the [card] distribution, we have agreed that:
1. A deck of cards has 52 cards divided in 4 equal[14] suits.
2. Each suit [symbolises] an aspect of the object (political, financial, social, and cultural).
3. Each card [symbolises] a rational argument in [pursuit of] the object on a specific aspect. In fact, we will turn the rational arguments […] into [numerical] values –cards– and risk them based on [the] rules of the Bridge game. These rules, as we have seen [in] Chapter [1.1], are a probabilistic combinatorial calculus, [where chance never has a] part.
4. The spades suit [symbolises] the political aspect [of the] object.
5. The hearts suit [symbolises] the financial aspect [of the] object.
6. The diamonds suit [symbolises] the social aspect [of the] object.
7. The clubs suit [symbolises] the cultural aspect [of the] object.
8. We have only one object to play: the occupied territories from the '67 War. [However], the territories are five: East Jerusalem, the Sinai Peninsula, the West Bank, Gaza and the Golan Heights (they are

---

[14] Equal in number –13 cards per suit– but not in score, [which is different for spades and hearts, and for diamonds and clubs]. [In a] Bridge [game] the score is important but, in our simulation, the suit score is [irrelevant] because in our case the importance comes from the argument related [to] the object that the card symbolises in each [particular] suit.

[mentioned here] in order of importance. This order will be used [...] for the card [assignment]).
9. The [card value is, in decreasing order]: Ace, K, Q, J, 10, 9, 8, 7, 6, 5, 4, 3, 2, [and] it is the [same] for each suit.
10. The [face value of each card] is: Ace=4 points; K=3 points; Q=2 points; J=1 point; for the [rest it] is null.
11. As we have 5 objects [...] and 13 cards [per] suit, [let us] agree that [...] every two consecutive cards [in each suit] will symbolise one object, according to the importance given [to the] objects, suits and cards, [as mentioned before]. [For instance, in our case] Ace and K of spades will symbolise the political arguments [for] East Jerusalem, Q and J [of] hearts will symbolise the financial arguments [for the] Sinai Peninsula, 10 and 9 [of] diamonds will symbolise the social arguments [for the] West Bank, 8 and 7 of clubs will symbolise the cultural arguments [for] Gaza, [...] 6 and 5 of [all suits] will symbolise the arguments [for] the Golan Heights.
12. Consequently, we will always have three surplus cards ([...] 4, 3 and 2 of each suit), [which] will be the added value [for] the generic [...] aspect [of each suit] (e.g.: [...] 4, 3 and 2 of spades will symbolise the added [generic] political value [for] the *object* [–the *object* being the whole] territories occupied from the *'67 War*).

Once we have agreed these 12 points, we can distribute the cards:

- ***Spades*** distribution (political arguments):

Ace: Political argument [for] **East Jerusalem.** [Face value] = **4 points.** [It goes to] **Begin** (it is the Jewish Holy City).

K: Political argument [for] **East Jerusalem.** [Face value] = **3 points.** [It goes to] **Sadat** (it is the Third Muslim Holy City and [it was an old Muslim territory of Jordan before the *'67 War*]).

Q: Political argument [for] the **Sinai Peninsula.** [Face value] = **2 points.** [It goes to] **Begin** (it is a war conquest and it entails the control [over one of the banks of the Suez Canal]).

J: Political argument [for] the **Sinai Peninsula.** [Face value] = **1 point.** [It goes to] **Sadat** (it [was an historical] Egyptian Territory and it entails the Egyptian control [over both banks of the Suez Canal]).

10: Political argument [for] the **West Bank.** [Face value] = **0 points.** [It goes to] **Begin** (it is a war conquest).

9: Political argument [for] the **West Bank.** [Face value] = **0 points.** [It goes to] **Sadat** (it [was an old Muslim territory from Jordan before] the *'67 War*).

8: Political argument [for] **Gaza.** [Face value] = **0 points.** [It goes to] **Begin** (it is a war conquest).

7: Political argument [for] **Gaza.** [Face value] = **0 points.** [It goes to] **Sadat** (it [was an old Egyptian territory before] the *'67 War*).

6: Political argument [for] the **Golan Heights.** [Face value] = **0 points.** [It goes to] **Begin** (it is a war conquest).

5: Political argument [for] the **Golan Heights.** [Face value] = **0 points.** [It goes to] **Sadat** (it [was an old Muslim territory from Syria before] to the *'67 War*).

4, 3 and 2: The added [generic political] value over the occupied territories. [Face value] = **0 points. 4:** [It goes to] **Begin** (he is the present 'landowner'); **3:** [It goes to] **Sadat** (the mediator, Carter offers it because he needs to have a foot in the *East World*); **2:** [It goes to] **Begin** (Carter is the ally of Begin, [thus he cannot] 'snatch' [the occupied territories from] Begin. [Besides], we should remember that Israel has nuclear weapons –manufactured at the nuclear plant in Dimona).

- *Hearts* distribution (financial arguments):

Ace: Financial argument [for] **Jerusalem East.** [Face value] = **4 points.** [It goes to] **Begin** (it is the Jewish and Christian Holy City. The religious tourism is a [source of] high financial potential [...]).

K: Financial argument [for] **East Jerusalem.** [Face value] = **3 points.** [It goes to] **Sadat** (it is the third Muslim Holy City. The religious tourism is a [source of] high potential financial [...], perhaps not so directly to Egypt but [it clearly is] to the Muslim world, [and particularly] to Jordan, the landowner of East Jerusalem before the *'67 War*).

Q: Financial argument [for] the **Sinai Peninsula.** [Face value] = **2 points.** [It goes to] **Begin** (it entails the control [over] one of the Suez Canal banks).

J: Financial argument [for] the **Sinai Peninsula.** [Face value] = **1 point.** [It goes to] **Sadat** (it [was] an old Egyptian territory [before] the *'67 War*. It entails the Egyptian control [over] the two Suez Canal banks).

10: Financial argument [for] the **West Bank.** [face value] = **0 points.** [It goes to] **Begin** (it could potentially contribute to increase the Gross Domestic Product).

9: Financial argument [for] the **West Bank.** [Face value] = **0 points.** [It goes to] **Begin** too (the territory could be developed economically but it would not be a benefit for Egypt as it did not belong to [it]).

8: Financial argument [for] **Gaza.** [Face value] = **0 points.** [It goes to] **Begin** (it could potentially contribute to increase the Gross Domestic Product).

7: Financial argument [for] **Gaza.** [Face value] = **0 points.** [It goes to] **Sadat** (it [was] an old Egyptian territory [before] the '67 War. It could potentially contribute to increase the Gross Domestic Product).

6: Financial argument [for] the **Golan Heights.** [Face value] = **0 points.** [It goes to] **Begin** (Israel needs the water from the Golan Heights).

5: Financial argument [for] the **Golan Heights.** [Face value] = **0 points.** [It goes also to] **Begin** (Egypt cannot have financial interest in a territory that did not belong to [it]).

4, 3 and 2: The added [generic financial] value over the occupied territories. [Face value] = **0 points. 4:** [It goes to] **Begin** (he is already the 'landowner'); **3:** [It goes to] **Sadat** (the mediator, Carter offers it because he needs the Egyptian oil); **2:** [It goes to] **Begin** (Carter is the ally of Begin, [so] he cannot betray Begin).

- *Diamonds* distribution (social arguments):

Ace: Social argument [for] **East Jerusalem.** [Face value] = **4 points.** [It goes to] **Begin** (it is the Jewish Holy City. It is the only one territory really occupied to the Jewish population).

K: Social argument [for] **East Jerusalem.** [Face value] = **3 points.** [It goes to] **Begin** ([as in the previous argument], the Jewish population is the majority in East Jerusalem).

Q: Social argument [for] the **Sinai Peninsula.** [Face value] = **2 points.** [It goes to] **Sadat** (the majority [of the] population is Egyptian. It [was] an old Egyptian territory [before] the '67 War).

J: Social argument [for] the **Sinai Peninsula.** [Face value] = **1 point.** [It goes to] **Begin** (it gives [Israel] the opportunity to increase population).

10: Social argument [for] the **West Bank.** [Face value] = **0 points.** [It goes to] **Sadat** (the majority of the population is Palestine).

9: Social argument [for] the **West Bank.** [Face value] = **0 points.** [It goes to] **Begin** (it gives [Israel] the opportunity to increase population).

8: Social argument [for] **Gaza.** [Face value] = **0 points.** [It goes to] **Sadat** (the majority [of the] population is Egyptian/Palestine. It [was] an old Egyptian territory [before] the *'67 War*).

7: Social argument [for] **Gaza.** [Face value] = **0 points.** [It goes to] **Begin** (it gives [Israel] the opportunity to increase population).

6: Social argument [for] the **Golan Heights.** [Face value] = **0 points.** [It goes to] **Sadat** (the majority of the population is Palestine and/or Muslim).

5: Social argument [for] the **Golan Heights.** [Face value] = **0 points.** [It goes to] **Begin** (it gives [Israel] the opportunity to increase population).

4, 3 and 2: The added [generic social] value over the occupied territories. [Face value] = **0 points. 4 and 3:** [It goes to] **Sadat** (the majority of the population in the occupied territories is Egyptian, Palestine and/or Muslim, [...] not Jewish); **2:** [It goes to] **Begin** (he is the present 'landowner').

- *Clubs* distribution (cultural arguments):

Ace: Cultural argument [for] **East Jerusalem.** [Face value] = **4 points.** [It goes to] **Begin** (it is the Jewish Holy City).

K: Cultural argument [for] **East Jerusalem.** [Face value] = **3 points.** [It goes to] **Sadat** (it is the third Muslim Holy City).

Q: Cultural argument [for] the **Sinai Peninsula.** [Face value] = **2 points.** [It goes to] **Sadat** (it is the memorable territory of Egypt).

J: Cultural argument [for] the **Sinai Peninsula.** [Face value] = **1 point.** [It goes to] **Sadat** (the Sinai is de facto a historical Egyptian territory).

10: Cultural argument [for] the **West Bank.** [Face value] = **0 points.** [It goes to] **Sadat** (the majority [of the] population is Palestine).

9: Cultural argument [for] the **West Bank.** [Face value] = **0 points.** [It goes to] **Begin** (it is located [within] the Promised Land border[15]).

---

[15] The *Promised Land border* taken from the *Holy Torah*, Genesis 15:18 (The Jewish Publication Society of America, 1917).

8: Cultural argument [for] **Gaza.** [Face value] = **0 points.** [It goes to] **Sadat** (the majority of the population is Egyptian/Palestine. It [was] an old Egyptian territory [before] the *'67 War*).

7: Cultural argument [for] **Gaza.** [Face value] = **0 points.** [It goes to] **Begin** (it is located [within] the Promised Land border).

6: Cultural argument [for] the **Golan Heights.** [Face value] = **0 points.** [It goes to] **Sadat** (the majority of the population is Palestine [and/or] Muslim).

5: Cultural argument [for] the **Golan Heights.** [Face value] = **0 points.** [It goes to] **Begin** (it is located [within] the Promised Land border).

4, 3 and 2: The added [generic cultural] value over the occupied territories. [Face value] = **0 points. 4 and 2:** [It goes to] **Sadat** (the majority of the population on the occupied territories is Egyptian, Palestine [...] or Muslim, [...] not Jewish); **3:** [It goes to] **Begin** (Four out of the five territories are [within] the Promised Land border).

Now we know the [card assignment], the result is as follows:

For Begin:
- ♠ Ace, Q, 10, 8, 6, 4, 2 = 7 cards/6 points in ♠
- ♥ Ace, Q, 10, 9, 8, 6, 5, 4, 2 = 9 cards/6 points in ♥
- ♦ Ace, K, J, 9, 7, 5, 2 = 7 cards/8 points in ♦
- ♣ Ace, 9, 7, 5, 3 = 5 cards/4 points in ♣

Total cards: C^{28}_{52} Total points: P^{24}_{40}

For Sadat:
- ♠ K, J, 9, 7, 5, 3 = 6 cards/4 points in ♠
- ♥ K, J, 7, 3 = 4 cards/4 points in ♥
- ♦ Q, 10, 8, 6, 4, 3 = 6 cards/2 points in ♦
- ♣ K, Q, J, 10, 8, 6, 4, 2 = 8 cards/6 points in ♣

Total cards: C^{24}_{52} Total points: P^{16}_{40}

[...] [Let us point out that, though the number of cards is not equal for both players], this is *not* a real problem because the difference can be considered [...] negligible, as we are [assuming] a balance of +/- 2 cards [per player] and [their] score in points [is] 24/40=60% for **Begin** and 16/40=40% for **Sadat**. This difference is acceptable because [...] at least one player [should have] 30% of the points; in our case there are two players (not four), [so one of them should have at least 60% of points, and Begin does].

[It is the time to know *which will be the trumps?*] […] It is evident that **Begin** will decide on the trumps (he has the majority, 60%, of the points at stake and in this case also [the majority of] the cards, [53.85%]). [In order] to play […] trumps [it] is necessary to have at least [one-half] + 1 of [the] possible cards on this suit (13/2=6.5≈7cards) and [it is] desirable [to have one-half] +2, [which] means 8 cards. **Begin** [will choose] to play at trumps in hearts, but **Sadat** [will choose] to play at trumps in clubs. As **Begin** is stronger [in] all suits, even in clubs, because he has the Ace + 4 cards to support it, he will decide that the trumps will be no-trumps. **Sadat** will agree to play at no-trumps. In the Bridge terminology, the [accord] is expressed as: 1♥ **Begin**, 2♣ **Sadat**, 3NT **Begin**, Pass **Sadat**, [Pass **Begin**]. The game starts and the promise is: "I **Begin**, [I] promise that playing at no-trumps, I will achieve at least 18 tricks [out of] the possible 24" [(18 tricks –because 3NT means 9 tricks when there are 4 players– [out of] 24 and not [out of] 26 because Egypt only has 24 cards)].

We are ready [for] the crucial moment of [the] game:
Begin

♠ Ace, Q, 10, 8, 6, 4, 2   ♥ Ace, Q, 10, 9, 8, 6, 5, 4, 2

♠ K, J, 9, 7, 5, 3   ♥ K, J, 7, 3

Sadat

                       Begin

                       ♦ Ace, K, J, 9, 7, 5, 2   ♣ Ace, 9, 7, 5, 3

                       ♦ Q, 10, 8, 6, 4, 3   ♣ K, Q, J, 10, 8, 6, 4, 2

                       Sadat

Sadat starts the game. At no-trumps [it] is convenient to play the fourth [card] of the long suit, or the highest [card of the long suit in the case of having a sequence; here, the second option is chosen]. Then **Sadat** will play K♣, since it is the only possibility to reinforce the Q♣ and J♣, because **Begin** will use the Ace in other suits to reinforce his position and if **Sadat** plays [any other suit] before **Begin**, he will always lose. The only possibility for **Sadat** is to agree with everything, although he could [remind **Begin**] his [own] political and financial force. [Facing this case], the only feasible options for **Sadat** are Q♣ and J♣. In a real Bridge game, **Sadat** could win other tricks [thanks to observance of turns to play, while in] a conversation, the turn is not [always respected].

Therefore, *what is the true profit for* **Sadat**? The true profit for **Sadat** is the cultural matter over the Sinai Peninsula (Q and J in clubs). *Where can* **Sadat** *[place] the strength in this negotiation?* **Sadat** can only put the strength on the importance of the return to the Sinai Peninsula. He will need to keep the negotiation around this matter all the time. That will be his key point. And, *what is the true profit for* **Begin**? The true profit for **Begin** is to reaffirm his power in the region.

At this time, [let us] remember the principle [that] should not be forgotten [...]: [the result of the game is not to be conditioned]. *What happened in Camp David?* The agreement was to return the Sinai Peninsula to Egypt. [The goal], *The Framework for Peace in the Middle East*, was achieved, and the *Framework for the Conclusion of a Peace Treaty between Egypt and Israel* was [agreed].

So, [we can now answer the questions]:

1. [*Is the Bridge game the element that bridges both spirals in negotiations?*]
   [...] Our forecast on *Camp David Accords* has been right. We can assert that the Camp David [summit was] a resounding success, [despite many people thinking that it was a failure]. According to the previous [reflection], the best rational decision is the intersection point between '*f(optimization)*' and '*f(satisfaction)*'. This point will be, at the same time, the [consequence] of optimizing to [the] maximum our true possibilities and *not* losing anything [but what was already lost] before the negotiation [...]. Herein the satisfaction: obtaining all obtainable and losing *only* the indispensable. To get more is a 'gift' [and to get less is] a 'loss', but [*neither* is the consequence of] a rational decision because [these] would [only] be the result of a mistake [during the] negotiation [process]. Therefore, we can say that Bridge bridges both spirals in [...] negotiations, because it [combines] optimization [and satisfaction (obtaining all what is obtainable and losing only what has already been lost before the negotiation)] with implementation. Bridge is *not* only a theory; it is a theory that can be implemented in reality.

2. [*Could the Bridge game be a tool for a rational analysis of negotiations?*]
   The Bridge game is a good application to analyse the real possibilities in a [...] negotiation, because it bridges both spirals [...], and because it [combines] optimization (obtaining all obtainable) and [...] satisfaction (losing only the indispensable, [what is already] lost before the negotiation) together with the implementation [which contains in itself the *win-win* idea]. [Besides], Bridge is *not* only a the-

ory, it is a theory that turns [...] [unquantifiable values] (the arguments) [into] countable [values] (the cards [with] their values), so we can play; and we can also [do] the other way round and [see] what has happened. Therefore, Bridge [can] be a tool for the rational analysis of [...] negotiations.

[We can now move on to go through the lights and shadows of the Bridge application on negotiations].

Chapter 1.3: The Bridge. A bridge toward Negotiations. Lights and Shadows

[This chapter is taken from Martínez-Cazalla, 2011, pp. 82-84. You will find in square brackets the additions and/or alterations to the original text].

As we announced at the [introductory pages of this act, this chapter is an attempt] to answer the following questions: *To what extent is it possible to consider Bridge as an element for a rational analysis of [...] negotiations? What light does it provide to our understanding of [...] negotiations? Which are its limitations?* [and] *How could we overcome these limitations?*

About the first question, *To what extent is it possible to consider Bridge as an element for a rational analysis of [...] negotiations?*, the answer has already been suggested at the conclusion of the previous chapter. [Firstly], we would like to [stress] the fact that Bridge is not only a theory, but a theory that turns [...] [unquantifiable values] (the arguments) [into] countable [values] (the cards [with] their values); [this is why] we can play, but we can also [do] the other way around, and see what has happened [in order to evaluate past negotiations, or to prepare future negotiations]. Bridge, connecting the two spirals has become an abducing tool, that is, Bridge makes [it] possible to propose the following syllogism: we have a great premise and it is evident, and we have [a lesser one, which is also] evident or only probable, [then Bridge allows us to link them together]. We have the theoretical spiral, which starts from a theoretical-deductive process, [which] is evident; and we have the implementation, the [...] negotiation itself, which starts from an empirical-inductive process because not all negotiation details are evident or probable. Finally, we need something to bridge, to link [theory and implementation, in order] to [reach] a conclusion. As Bridge makes [it] possible to turn [the unquantifiable values of the negotiation] (that is, the empirical values) into countable [values] (the theoretical values), and inversely, then we can find the link between both. We are now ready to arrive to a conclusion. Bridge [meets the two points of view that Rudnianski & Bestougeff (2007) pointed out in their work *Bridging Games and Diplomacy*]: "(...) prescriptive studies [theoretical-deductive], concerned with procedures for achieving good outcomes, and the descriptive studies [empirical-inductive], focused on understanding how people negotiate" (p. 150). [...] Although Bridge belongs to the *Game Theory*, it offers the possibility of *not* being only a laboratory analysis. In this way, the Bridge application is a step forward [from] the *Adjusted Winner Procedure* (cf. Brams & Taylor, 1996, pp. 68-75), [because] in Bridge we [not only] ascribe a value [to] the different aspects of the issue, [which] is the same starting point [as that of] the *Adjusted Winner Procedure*, [but] Bridge puts these values [–issues–] at risk [for what they are worth by] themselves, and simultaneously [for what they mean for the players and for their goal]. The value [fluctuates] from virtual to real [and] it is a constant feedback. [Therefore, Bridge allows for having a goal beyond the object at stake].

In order to answer the question: *What light does [Bridge] provide to our understanding of [...] negotiations?* [...] Bridge gives us the opportunity to [corroborate whether] a [certain] [...] negotiation was, or will be, a success or a failure. As [a] linkage element, it [yields] simultaneously an overview and an internal detailed vision. We [...] need to remember now [...] that the winner is the pair that has reached 100% of the possible tricks [in more hands]. Therefore, the Bridge application [tells] us about the degree of true success or true failure: [a non-random success or failure, which is just a mathematical result, and thus, a *win-win* result]. [We often] conclude: "the negotiation was or will be a success" or "the negotiation was or will be a failure", but [many times] we mistake our wishes and/or fears [for] our real possibilities, and we assess events incorrectly. The right [assessment of] events is the only sure way toward the future, and it is [a means of learning for future events]. It [gives us] the possibility to accept to negotiate or not, in [sight] of our cards and our real chances to win tricks with them.

Finally, *Which are the limits of Bridge?*, and *How could we [overcome] these limitations?*
Bridge has [...] limitations [that cannot be circumvented, whenever] there is [...] [one or a combination of the following circumstances]:
- The [object/s to negotiate do] not agree [...] [with the aim/s of the negotiation. Application of Bridge requires that object/s and goal/s have the same internal structure, so that the same suits must represent their different aspects. An example of the opposite is: if we need to negotiate the price –money– to reach something different from

just more money, the issue and [the] objective [do] not have the same internal structure. On the contrary, if the case is negotiating the price —money— for a financial product, object and objective [...] have the same internal structure].

- The power is very close to [one of the players]. In this case [...] the best solution [could] be the *Adjusted Winner Procedure*, [since it] calculates issues and *not* relationships between [those] and the players.

- [...] Control over [...] time and [...] space. All changes [will alter the game], even [small ones]. [...] An unexpected event [in] the last minute [may change it all after a very good preparation], even if we have done the calculus of gains and losses in connection with the different combinatorial possibilities. ([...] We should [prepare and] make [all the calculi] before getting into the negotiation [...] because it is the only possibility to know how often [a] favourable scenario for negotiating [will] take place). [Unexpected events are] much more common [than we] think, because [...] negotiations are implemented [by] individuals, and individuals are [*not* 100% predictable]. This could be [overcome] with a [thorough] preparation [knowing that some uncertainty will always remain —subjectivity is always there, it is intrinsic to the players, subjects themselves]. Briefly, Bridge applied [...] [on] negotiations is an analysis about [*whether the arguments we have can be tackled in the negotiation* and about *when to say what?*] In case of doubt, we [should keep in mind this]: "Whereof one cannot speak, thereof one must be silent" (Wittgenstein, 1922, para. 7).

[...]
[The great limitation of the Bridge application on negotiations is, and will always be, subjectivity, which is present from the beginning of the process and is liable for two direct shadows, inherent to our own subjectivity, and for one indirect shadow, implicit in our interlocutor].

[Direct shadows]:

[1. To choose what arguments should be employed during the negotiation].

[2. To assign to each argument the right card to symbolise it].

[Indirect shadow]:

[1. Even when trying to be as conscientious as possible during the negotiation process, we cannot guarantee what will be replied to our arguments, we cannot predict that].

[The next pages are meant to throw some light on each of the above shadows. The negotiator, as subject with subjectivity, is one of the necessary elements in a negotiation, so Shadows will definitely persist but they become lighter].

ACT II: Neuro-Linguistic Patterns (NLP)

As mentioned in the introductory pages, each act aims to be a part of a tragedy: a 'conceptual mimesis' of the used arguments, meant to produce a 'catharsis', an improvement in the negotiation process. Thus, this act is our second step where our task will address to the limits, with the purpose to reduce them, to throw some light on the shadows.

At this step the tool chosen to try to reduce the limitations has been Neuro-Linguistic Patterns (hereunder, NLP); NLP will be applied to the previously showed shadows. NLP have been selected because all the flaws in the negotiation process are a product of the negotiators' weakness shown through their arguments. If we want to study flaws in arguments, we are in the field of NLP.

Before starting this act, we would like to give two important remarks to tackle it:
1. We have just said above *Neuro-Linguistic Patterns* and *not Neuro-Linguistic Programming*. Thus, when we talk about NLP we will *only* be talking about *Patterns* and *not* about Programs; a Program is just the effect of applying a Pattern. When we expound on an idea, this is always put into words, that is, in a formal grammar structure (Pattern); this structure will produce a reaction, and at this moment the Pattern becomes a Program, as result of its implementation. The better we know Patterns –causes–, the better control we can have over unleashed Programs –effects–, that is, replies to the presented arguments –both the arguments and their replies can only be expressed in a Pattern. Nevertheless, subjectivity, so, unpredictability, will always remain.
2. We are *only* analysing Patterns, as the most strictly formal grammar structure, in order to know the natural unleashed Programs and to develop some sort of control of the possible replies. Patterns, like operative system foundations, are quite few and specific. Therefore, the length and depth of this act is restricted accordingly and includes only a handful of essential references.

In the next pages, we will try to give an answer to the question: *What can NLP possibly do to illuminate the limitations?* Firstly, we will introduce NLP in order to decide whether they may be considered a useful tool to make rational decisions in the world of negotiations –Chapter 2.1. Right before we deal with limitations under the light of NLP we will look at how are these Patterns operating in our mind –Chapter 2.2. To finish, we will look at the shadows under the light of NLP –Chapter 2.3.

Chapter 2.1: Are NLP a useful tool to make rational decisions in the world of negotiations?

In order to give an answer to this question, we should start by defining NLP:

> Neuro-Linguistic Programming (…) is the study of [Neuro-Linguistic Patterns, of] how we think and communicate, with ourselves and with others, and how we can use this to get the results we want. (Centre of Excellence, 2014a, p. 11)

Therefore, reading this definition we can trust NLP are the right tool to reduce the limitations. However, when we go deeper in we discover that NLP are more about Programming than about Patterns. It is more focused on the techniques to put into practice and the results than on the structure of arguments themselves. Moreover, since its beginning, NLP focused in Programming; that is why there is a large amount of literature on NLP (*P* understood as *Programming*) applied on business (*business* meaning *negotiation per se*). *This is* **not** *the interest of this study. This piece of research is* **not** *about how to say an argument* (Programming) –in a semiotic meaning–; *we are analysing how to build it*, its structure –syntax– and the semantic implication of the syntax chosen (syntactic Pattern form) in addition to how it is taking place – intonation– (pragmatic Pattern form), and what is behind the manifested form, what is hidden, as if modelling a precise syntax and its intonation for each argument.

Now, we can agree that NLP (*P* understood as *Patterns*) are a useful tool to make decisions in the world of negotiations. *Could NLP also be useful to make rational decisions?* Yes, definitely, because we are keeping the *P* in NLP as Patterns, i.e., as syntax + 'semantics of syntax in itself'[16], as *grammar* in Montague's meaning (cf. Montague, 1970, pp. 373-398), that is, we could

---

[16] We are understanding *semantics of syntax in itself* as the representational value that the syntactic form takes in our mind (e.g.: when we hear a conjunctive sentence, our mind considers every part of the conjunctive sentence is the case, meaning *true*).

consider Patterns as a Montague's subset. Moreover, we know that the rational truth-value is in syntax, in accordance with:
- If an elementary proposition is true, its negative mood will be false (taking the same sense as in classic Logic).
- For a true conjunctive sentence, each part of it must be true (taking the same sense as in classic Logic).
- For a true disjunctive sentence, at least one of its parts must be true (taking the same sense as in classic Logic).
- For a true conditional sentence, we consider *only* the case our mind takes as natural: if the antecedent is true –assumed to occur–, then the consequent will be also true.
- Any true universal sentence or necessary sentence must be true for every case (taking the same sense as in classic Logic).
- Any true particular (existential) sentence or possible sentence must be true at least for one case (taking the same sense as in classic Logic).

If we are to be true as negotiators, then our sentences will be true sentences. Therefore, their syntax will be in accordance with what has been said above. Thus, we can conclude that NLP are a useful tool to make rational decisions in the world of negotiations.

Now we are almost ready to throw some light on the limitations suggested at the end of the previous act. Right before engaging in limitations, we will look at how these Patterns work in our mind, what kind of Programs they are triggering.

Chapter 2.2: How do Linguistic Patterns work in our mind?[17]

Linguistic Patterns, syntactic expressions of thoughts, operate in two directions:
- *Toward us*: When somebody is talking to us, we are doing linguistic assumptions. These assumptions are creating a new syntactic structure (the same or different from the original), meaning a new truth-value, which could also be the same or different from the original truth-value in the mind of the speaker. This new truth-value is the one to which we will reply. As we cannot have control of the original meaning in the mind of our interlocutor, we cannot be sure that

---

[17] Cf. Centre of Excellence, 2014a, pp. 55-71.

our answer is really in accordance with what they[18] have said. Nevertheless, we can learn a great lesson from this process: the same happens the other way round.
- *Toward our interlocutor.* We cannot control the structure of our interlocutor's speech but we can have a good command over our language expression. To reach this command we need to assess presuppositions, or rather, linguistic assumptions; they are basically:

Syntactic Patterns
- Using elementary propositions presupposes the existence of the entity to which they are applied. [Elementary proposition, expressed by the symbol A]
- Using conjunctive sentences presupposes that each one of them is the case. [Conjunctive sentence, $\alpha \wedge \beta$]
- Using *or* presupposes exclusion/inclusion. [Disjunctive sentence, $\alpha \vee \beta$]
- Using the conditional structure, like if … then … or similar, presupposes a cause and its effect. In fact, every cause–effect can be represented as a conditional relationship. This is how our mind understands *viscerally* the conditional pattern. Even when *not* every conditional sentence is a proper representation of the cause–effect relationship, *a conditional sentence will produce a cause–effect pattern in our mind, that is, the antecedent is assumed to occur.* [Conditional sentence, $\alpha \rightarrow \beta$]
- Using quantifiers or modal operators presupposes necessity or possibility. [Universal, $\forall_x A_x$ /Necessary, $\square A_{ci}$ sentence; or Particular, $\exists_x A_x$ /Possible, $\lozenge A_{ci}$ sentence]
- Using negative sentences that are true, their affirmative mood will be false. But WARNING! Negative sentences that are a command –more or less explicit–, are tricky for our mind, which acts quite oddly: "(…) the mind cannot process a negative instruction, for example, if I say don't think of the colour red, what comes to mind?" (Petruzzi, 2012, p. 77). Therefore, "Suggestions which logically are negatives but create an internal representation which the unconscious acts on as a positive suggestion" (Centre of Excellence, 2014a, p. 69). Negative sentences under a ques-

---

[18] For the sake of inclusive language, all 3rd person singular pronouns whose antecedent is a genderless person have been changed into the plural form (*he* and *she* for *they*, *him* and *her* for *them*, *his* and *her* for *their* and *his* and *hers* for *theirs*).

tion form are also equivocal for our mind; in this case, our mind does not know whether the right answer is the affirmation for the negative question or its negative mood. [Negative sentence, $\neg\alpha$]

Intonation Patterns (pragmatic Pattern form)
- Question [?] / Statement [!] / Command [!]. An argument is not only a sentence; an argument is a pronounced sentence. Thus, intonation is as important as the argument itself.

Special Pattern
- Silence: "(...) silence can be interpreted as having meaning. (...) Our mind and body are part of the same system, so the thoughts we have affect our physiology, and it shows up in our non-verbal interactions" (Centre of Excellence, 2014a, p. 20). Moreover, silence is the replacement of at least a word, or even the expression of a whole thought. We cannot be so naïve as to believe that silence is an empty set or an infinite set; silence is always the expression of at least one of the elements of a limited set of possible elements to take up that place, because silence encodes usually a finite number of replies to the question that triggered the silence.

So far, we have been speaking about the visible structure of Patterns. Patterns, we should realise, need to take place in a concrete expression (word/sentence + intonation + body expression, or silence + body expression); this is why we cannot overlook what is behind every concrete expression. In the farthest depths of every one of us are our beliefs and values – hidden in the structure of the Patterns–; they are modelling our expressions, *not* only their content *but* also their form –syntax + intonation. Being aware of how beliefs and values are modelling expression forms (direct/indirect syntactic patterns/intonations/body language) is not easy (for instance, oriental cultures would usually choose the indirect form), but it is the essential key to achieve rapport, so we should *not* disdain our interlocutor's beliefs and values when evaluating the patterns that we are going to use. When communicating with somebody, we may know and/or share beliefs and values with them or not, but if we really want to communicate we will have to shape our expressions in a common territory to guarantee comprehension, to make sure our patterns are recognised.

Knowing all that, we can make as many combinations, variations and/or permutations as we want. We can say, being right, that if we are aware of this, we will manage our arguments as good as possible in the desired way; even when we "(…) don't get dealt the best cards in life, though we have a choice in how we respond. And it is within our response that our lives are shaped" (Petruzzi, 2012, p. 29). Besides, the desired way cannot be manipulative because "(…) manipulative behaviour never works. Usually the other person will spot what the manipulator is trying to do because they seem 'false' or not fully human in their responses. (…) If you respect the other person and dovetail your desired outcome with theirs, you will get a *win-win* situation and everyone is satisfied. If not, neither party will achieve their outcome" (Centre of Excellence, 2014a, p. 29).

Now it is the time to see whether it is possible to throw some light on the limitations suggested at the end of the previous act.

Chapter 2.3: Shadows under the light of NLP

Here we go back to the question of the great limitation for the Bridge application on negotiations, as mentioned at the end of the previous act in Chapter 1.3, and look at it under NLP: *Can NLP throw any light on the shadows?* We will go through the limitations one by one.

We will start by looking at the ones we named *direct shadows*, the ones inherent to our own subjectivity.

1. To choose what arguments should be employed during the negotiation.

    As discussed earlier, we cannot be sure about what arguments, in the meaning of their content, should we choose. Nevertheless, we are responsible of choosing the best syntax, intonation and body language pattern, or the best silence and body language pattern, for each of the arguments. To do our best, we should always keep in mind where is the common territory of beliefs and values (our own and the interlocutor's) and we should make the arguments be born there.

    At this moment, there is no doubt about a couple of things we have to take into account when we choose arguments to employ in a negotiation process:

- Choosing arguments under a conditional form produces a cause–effect pattern in our mind, that is, the antecedent is assumed to occur.
- When choosing arguments in negative mood, one must be extremely careful (as expounded in pages 30-31):
    - In case they enclose a command: they will suggest the opposite –a positive order– in the mind of our interlocutor.
    - In case they are expressed as a question: our interlocutor will feel confused to answer it.

2. To assign to each argument the right card to symbolise it.

We will never be one hundred per cent certain of having chosen the right card, the one that best symbolises each argument. However, the better our knowledge of the common territory of beliefs and values, the better the connection, and therefore also the accuracy, between each argument and its symbol in a playing card.

Regarding the *indirect shadow*, the one implicit in the subjectivity of our interlocutor, we cannot guarantee what will be replied to our arguments, we cannot predict it, so in this regard we may say that no light is coming in from NLP. We could only suspect that the replied arguments will probably come from the interlocutor's beliefs and values, but we cannot be sure. In spite of that, a small light is on sight: we could suspect, with a high degree of certainty, which will be the formal grammar structure –syntax– chosen for each reply as result of the Program triggered for the previous argument expressed in a Pattern. Thus, to be attentive to building *good arguments* (arguments expressed in a good 'container expression form' –pattern–) is extremely relevant. As has been shown, we can have much more control over the form than over its content, then our arguments need to be *valid arguments* (containing expressions of pertinent ideas), and they must also be *good arguments* –pertinent ideas 'wrapped' in a good pattern.

Next act will be an attempt to pursue the first *direct shadow* and the *indirect shadow*. For both, we will be going deeper in syntax, because syntax is the *only* rational pattern for an argument; therefore, syntax is the *only* discipline where we can predict the outcome and here we should do our best.

ACT III: Cryptic Language: *Russian Cards*

As promised at the end of the previous act, in this one we will tackle deeper the syntactic patterns as a tool to decide what arguments should be chosen to employ during a negotiation, as well as to predict replies, both still in the darkness.

Along the next pages, we will try to give an answer to *how to create right syntactic patterns which can cause good and predictable replies*. Hence, we will analyse a cryptic language: *Russian Cards*. This language was developed by van Ditmarsch (Van Ditmarsch, van der Hoek, & Kooi, 2008, pp. 97-104 et 108) –Chapter 3.1. We will extract two key lessons from it –Chapter 3.2. Finally, a dialogical framework will throw some light on the darkness –Chapter 3.3.

Chapter 3.1: Hunting Hackers. Is it useful to think like a 'mute guest' to prepare a negotiation talk thoroughly?

This chapter deals with a cryptic language: *Russian Cards*. The original problem was proposed at the Moscow Mathematics Olympiad in 2000:

> **Level C. Problem 6.** Seven cards were drawn from a deck, shown to everybody, and shuffled. Then Greg and Linda were given three cards each, and the remaining card was either (a) hidden or (b) given to Pat. Greg and Linda take turns announcing information about their cards. Are they able to ultimately reveal their cards to each other in such a way that Pat cannot deduce the location of any card he doesn't see? (No special code was set up in advance; all announcements are in "plain text"). (Fedorov, Belov, Kovaldzhi, & Yashchenko, 2011, p. 5)

Thus we are inside a typical framework for a Dynamic Epistemic Logic –DEL. The progress of knowledge depends on the public announcements (cf. Van Ditmarsch et al., 2008, pp. 104-107) made by the knowledge subjects involved –in our case: Greg, Linda and Pat. We can say Greg and Linda are active knowledge subjects, and Pat is a passive one, like a 'mute guest'.

Is the 'mute guest' a true passive knowledge subject or could he be a 'hacker in disguise'? The aim of this chapter is to answer this question. We will analyse the *Russian Cards* from Dialogical Semantics because our interest is to find out: *What is happening in Pat's mind? What is Pat thinking when he is listening to an-*

nouncements from Greg and Linda? Notice that we are *not yet* using the word *knowledge* for Pat, because the problem states: "Greg and Linda can exchange information about the hands they hold without Pat being able to deduce the owner of any card other than his own". Our interest here is in the field of semantics, in the meaning of what is being said. Only at the end of this reasoning, we will get to know what has changed in Pat's knowledge.

For this analysis, we will take the *Russian Cards* problem developed by van Ditmarsch (Van Ditmarsch et al., 2008, pp. 97-104 et 108). First, we will tackle its dialogical semantics form for the general case. Then we will be ready to think what is happening in Pat's mind in each one of the shown cases. Note that in the van Ditmarsch's *Russian Cards* the names of the characters have been changed: Greg is Ann (a), Linda is Bill (b) and Pat becomes Cath (c).

A Dialogical Semantics[19] for the *Russian Cards*[20]
1. Mathematical rules for the *Russian Cards*
1.1. Characteristics of the game:
- We have three players: a; b; c.
- We have a stack with seven different cards. They are numbered: 0; 1; 2; 3; 4; 5; 6.
  $C = \{0, 1, 2, 3, 4, 5, 6\}$
- The card deal for player 'a' and for player 'b' is the same: three cards each; player 'c' gets only one card.
- The language assumes expressions in the form gR(m,n,p), that should be interpreted as player 'g' holds cards m; n; p. More precisely: aR(m, n, p); bR(m', n', p'); cR(m'') where m; n; p; m'; n'; p'; m" are numbers –not repeated– from 0 to 6.

   In order to simplify the notation, we will follow the convention used by van Ditmarsch, from whom expressions of the form aR(m, n, p) are taken: mnp$_a$ and so on.

1.2. Objective of the game:
The game has two parts and one objective:
Part 1: Players 'a' and 'b' interchange information about the cards they hold.

---

[19] Dialogical Semantics in the *intuitionistic* framework. For rules of intuitionistic Dialogical Semantics –points 2, 3, 4 and 5 of this section– cf. Rahman & Clerbout (2015) and Redmond & Fontaine (2011).
[20] Fulfilling *correction* and *completeness* within the Dialogical Epistemic Multi Agent Logic (DE-MAL) framework. Cf. Magnier (2013), pp. 80-98.

Part 2: After this sharing of information, player 'c' must be still ignorant, or in other words, they still know only their own card and do not know who has what.

The language assumes expressions in the form $K_g(mnp_g)$, that should be interpreted as player 'g' Knows that player 'g' holds cards m; n; p. Hence the objective of the game can be rendered with the following expression:

$[K_a(mnp_a) \wedge K_a(m'n'p'_b)] \wedge [K_b(m'n'p'_b) \wedge K_b(mnp_a)] \wedge [K_c(m''_c) \wedge \neg K_c(mnp_a) \wedge \wedge \neg K_c(m'n'p'_b)]$ that reads.

1.3. Knowledge stage or terms of the game:

- The 3 players (a; b; c) know that 7 cards have been dealt. They are *not* duplicated and they are numbered 0 to 6: C = {0, 1,.., 6}
- The deal has been: $C_{(a)}^3{}_7 * C_{(b)}^3{}_4 * C_{(c)}^1{}_1 = aR(^3{}_7); bR(^3{}_4); cR(^1{}_1) = 140$ deals are possible.
- After the deal the cards distribution has been: 012_a; 345_b; 6_c.
- At first, players know only their own cards.
- Player 'a' and player 'b' have to let each other know the cards they hold without discovering them to player 'c'. Player 'c' has to remain ignorant about who has what after their announcements (in accordance with the objective of the game –previous section: 1.2.– and inside the framework of the Logic of public announcements).

2. Particle rules (symbols taken as expressed in pages 30-31):

Announcement structure	Attack	Defence
!α∧β The attacker chooses the defence	?L∧ (Left side of the conjunction)	!α
	?R∧ (Right side of the conjunction)	!β
!α∨β The defender chooses the defence	?∨	!α
		!β
!α→β	!α (α is assumed to occur)	!β

$!\neg\alpha$	$!\alpha$	---------- (No defence, *only* counterattack is possible)
$!\forall_x A_x$	$?k$ (k is chosen by the attacker)	$!A_k$
$!\exists_x A_x$	$?\exists$ (could you show me one, please?)	$!A_k$ (k is chosen by the defender)
$!\Box A_{ci}$	$?c_j$ <ciRcj> (cj is chosen by the attacker)	$!A_{cj}$
$!\Diamond A_{ci}$	$?\Diamond$ (could you show me a case, please?)	$!A_{cj}$ <ciRcj> (cj is chosen by the defender)

Note: '$\Box$' and '$\Diamond$' will operate in the same way in all the cases where there is a modal operator: alethic, deontic, epistemic, doxastic, temporal or a combination of them.

3. Structural rules for a game played in the intuitionistic Logic framework:

- Player 'c' always remains as 'mute guest'.
- The game starts with an assertion from player 'a'.
- By rotating turns, player 'a' first, then 'b', and again 'a' and 'b', make a public announcement, either as an assertion or as a question.
- Each announcement –assertion or question– must be true. Each question must refer to a previous announcement.
- Each announcement produces a new engagement that adds to the previous ones, making an engagement chain. No player can avoid their engagement chain.
- No player can repeat an argument already attacked. If an argument is repeated, it will be because the player arrives to the same argument through a different way, for instance, from another hypothesis.
- Each announcement has to have a reply. It is not possible to leave an announcement without reply. At the end of the game each attack must be completed with its defence, unless:
    - The attack has been against a negative sentence. Then, no reply, no defence, is possible.
    Cf. Rahman & Clerbout, 2015, p. 68.

	O			P	
				! $A \vee \neg A$	0
1	? [$A \vee \neg A$]	0		! $\neg A$	2
3	! A	2		----	

O Wins

- The attack has been a double negative sentence. Negative sentences can only be attacked one at a time because, as already mentioned, no player can avoid their engagement chain, so no player can say !A when they have already said !$\neg A$ (being A an elementary proposition). Therefore, faced with attacks constructed with double negative sentences, a double attack is not possible.

Cf. Rahman & Clerbout, 2015, p. 69.

	O			P	
				! $\neg\neg A \rightarrow A$	0
1	! $\neg\neg A$	0		--	
	--		1	! $\neg A$	2
3	! A	2		----	

O Wins

- The attack has been an elementary proposition and the respondent does not have the possibility to reply the same elementary proposition.

Cf. Rahman & Clerbout, 2015, p. 66.

	O			P	
				! $Qa \rightarrow Qb$	0
1	! Qa	0		----	

O Wins

- *The best defence is a good attack.* If we can choose between attacking and defending, in most instances we should attack first.
- The game ends when 'a' knows b's cards and vice versa, and 'c' remains ignorant.

4. Formalisation for the general case:

Case 3.1

OPPONENT (b)			PROPONENT (a)		
HYPOTHESIS			**THESIS**		
H1	$g \neq g' \neq c$; $g, g' \in \{a, b\}$		$!(012_a \vee 012_b) \rightarrow [(012_a \vee 012_b) \wedge (345_a \vee 345_b)]$		0
H2	$C=\{0,1,2,3,4,5,6\}$				
H3	$mnp_g \rightarrow (\neg mnp_g \vee \neg m'n'm''_g \vee \neg n'p'm''_g \vee \neg m'p'm''_g \vee m'n'p'_g)$				
H4	$\{m \neq n \neq p \neq m' \neq n' \neq p' \neq m''\} \in C$				
H5	$m'' \in \{c\}$ then m''_c				
1	$!(012_a \vee 012_b)$	0		$![(012_a \vee 012_b) \wedge (345_a \vee 345_b)]$	4
3	$!012_a$		1	?V	2
5	$?L_\wedge$	4		$!(012_a \vee 012_b)$	6
7	?V	6		$!012_a$	8
9	$?R_\wedge$	4		$!(345_a \vee 345_b)$	10
11	?V	10		$!345_b$	20
13	$![012_g \rightarrow (\neg 012_g \vee \neg 346_g \vee \neg 456_g \vee \neg 356_g \vee 345_g)]$		(H4; 3) H3	!m/0; n/1; p/2; m'/3; n'/4; p'/5; m''/6	12
15	$![012_a \rightarrow (\neg 012_b \vee \neg 346_b \vee \neg 456_b \vee \neg 356_b \vee 345_b)]$		(H1) 13	!g/a; g'/b	14
17	$!(\neg 012_b \vee \neg 346_b \vee \neg 456_b \vee \neg 356_b \vee 345_b)$		15	$!012_a$	16
19	$!345_b$		17	?V	18

Summary for case 3.1: 'a' holds 012 and 'b' holds 345

Or:

Case 3.2

OPPONENT (b) HYPOTHESIS				PROPONENT (a) THESIS		
H1	$g \neq g' \neq c;\ g, g' \in \{a, b\}$			$!(012_a \vee 012_b) \rightarrow [(012_a \vee 012_b) \wedge (345_a \vee 345_b)]$		0
H2	$C = \{0,1,2,3,4,5,6\}$					
H3	$mnp_{g'} \rightarrow (\neg mnp_g \vee \neg m'n'm''_g \vee \vee \neg n'p'm''_g \vee \neg m'p'm''_g \vee m'n'p'_g)$					
H4	$\{m \neq n \neq p \neq m' \neq n' \neq p' \neq m''\} \in C$					
H5	$m'' \in \{c\}$ then m''_c					
1	$!(012_a \vee 012_b)$	0		$![(012_a \vee 012_b) \wedge (345_a \vee 345_b)]$		4
3	$!012_b$		1	?V		2
5	$?L_\wedge$	4		$!(012_a \vee 012_b)$		6
7	?V	6		$!012_b$		8
9	$?R_\wedge$	4		$!(345_a \vee 345_b)$		10
11	?V	10		$!345_a$		20
13	$![012_{g'} \rightarrow (\neg 012_g \vee \neg 346_g \vee \vee \neg 456_g \vee \neg 356_g \vee 345_g)]$		(H4; 3) H3	$!m/0;\ n/1;\ p/2;\ m'/3;\ n'/4;\ p'/5;\ m''/6$		12
15	$![012_b \rightarrow (\neg 012_a \vee \neg 346_a \vee \vee \neg 456_a \vee \neg 356_a \vee 345_a)]$		(H1) 13	$!g/a;\ g'/b$		14
17	$!(\neg 012_a \vee \neg 346_a \vee \neg 456_a \vee \vee \neg 356_a \vee 345_a)$		15	$!012_b$		16
19	$!345_a$		17	?V		18

Summary for case 3.2: 'a' holds 345 and 'b' holds 012

5. Interpretation keys:

- External columns contain the intervention order, that is, the number of game moves.
- The number of move that is being attacked is placed in the internal double column. Numbers on the left mean the opponent is attacking a move from the proponent [e.g.: move 1 (opponent) is attacking move 0 (proponent)]; numbers on the right mean the proponent is attacking a move from the opponent [e.g.: move 2 (proponent) is

attacking move 1 (opponent)]. Numbers in brackets above the attacked move number mean: "Based on what you said in move x or in your hypothesis y, I can attack you, as I am doing now" [e.g.: move 14 (proponent) is attacking move 13 (opponent), their attack based on Hypothesis 1 (H1)].
- The central columns contain announcements: centre left are the opponent's announcements and centre right are the proponent's.
- Each announcement is preceded by a sign:
  !: This means that the announcement is an assertion.
  Assertions are the pragmatic form of an attack or of a defence.
  ?: This means that the announcement is a question about a previous announcement and therefore, it does not introduce into the dialogue any new datum of knowledge.
  Questions are the pragmatic form of an attack.
- Each row comprises 6 boxes (from left to right):
  1. Box for the number of the opponent's move.
  2. Box for the opponent's announcement (attack or defence under the form of a question or an assertion).
  3. Box for the number of move attacked by the opponent to the proponent –if this is the case.
  4. Box for the number of move attacked by the proponent to the opponent –if this is the case.
  5. Box for the proponent's announcement (attack or defence under the form of a question or an assertion).
  6. Box for the number of the proponent's move.
- Box 2 and box 5 must be coordinated: if box 2 is an attack, then box 5 must be its defence (which does not necessarily have to be the next move). Thus, we will have an attack and its defence on the same row, and it is not relevant if the defence is the next move or if it happens many moves after the attack (e.g.: move 11 is an attack by the opponent to move 10 of the proponent. This attack is defended –replayed– on proponent's move 20). Therefore, each new attack must be placed in a new row in order to keep its defence box empty.

6. Practical cases:

Once we have dealt with the semantics for the general case, we should tackle the cases proposed by van Ditmarsch one by one, following the convention used by himself (Exercise 4.72 ff., p. 103). This will help us answer the question we proposed: *Is the 'mute guest' a true passive knowledge subject or could they be a 'hacker in disguise'?* To accomplish this we are *not* going to formalise each case in its dialogical semantics form. There is no need to repeat it for each case, once we know how Dialogical Semantics works, because our interest is in the hidden column, the one for the 'mute guest', Cath (c). Therefore, we will hear the announcements as 'c' would listen to them and we will imagine what kind of reflections would be happening in her mind.

Exercise 4.72 (A five hand solution): Assume deal of cards 012.345.6. Show that the following is a solution: Anne announces: "I have one of {012, 034, 056, 135, 246}" and Bill announces "Cath has card 6".

Bill (b) opponent		Ann (a) proponent		Cath (c) 'mute guest'
2	!6_c	!(012_aV034_aV056_aV135_aV246_a)	1	???

What is 'c' thinking after a's and b's announcements?
1. How did 'b' know I have card 6?
2. If 'b' said 6 and no other out of the four possible cards, those ones that 'b' does not have, must be because:
   2.1. In triads announced by 'a' where card 6 is there must be also at least one of b's cards so that 'b' can recognise these triads as *not* a's triads. Therefore, 'a' will never have any binomial accompanying card 6.
   2.2. Besides, each one of the triads mentioned in 2.1. should also guarantee the safety of the announcement, since no player should be inside the 'truth zone' of another one (you can only have total control of your 'truth zone'). To be sure *not* to trespass another player's 'truth zone', the only possible solution, for our case, is to include *only one* real card number of the announcer in each one of these triads.
   2.3. 'c' summarises her reasoning: In the triads where 6 is, there is also a card from 'b' (as seen in 2.1.). Thanks to this, 'b' knows these triads as *not* a's triads. These triads contain also one of a's cards to guarantee the safety of the announcement (as seen in 2.2.).
3. Thus, 'c' removes these triads from a's announcement, and the result is a 'new' a's announcement in c's mind: !(012_aV034_aV135_a).
4. 'c' asks herself what is a's hand. To answer, she will be doing the following reasoning:

4.1. In the triads where 6 is (056; 246), there is also one card of 'a' and another of 'b', so:

a = 0; 2   then   b = 5; 4
a = 5; 4   then   b = 0; 2
a = 0; 4   then   b = 5; 2
a = 5; 2   then   b = 0; 4

4.2. Next step in c's reasoning is comparing these binomials to the three triads remaining in c's mind (as seen in point 3): $012_a \vee 034_a \vee 135_a$. The result is that either 'a' holds **012** or **034**, therefore 'a' holds 0 and 'b' must hold either 54 or 52, so 'b' holds 5. None of these binomials allows 'a' to hold 135, so this triad is no more an option. Therefore, 'c' knows 'a' must have $012_a \vee 034_a$.

4.3. Now, 'c' compares the possible a's binomials to the triads containing card 6 (056; 246) looking for more information.

012 to 056: as 'b' said 'c' holds 6, and 'c' knows already 'b' holds 5 and 'a' holds 0.

034 to 056: This is the same case as above.

012 to 246: as 'b' said 'c' holds 6, and 'c' knows 'a' holds 0, and 0 should be together with 2 or with 4. For this case 'c' would think 'a' holds 2 and 'b' holds 4.

034 to 246: This is the opposite of the previous case. Here 'c' would think 'a' holds 4 and 'b' holds 2.

After this reasoning, we can conclude that nothing new can be known. 'c' arrives to the same conclusion as in the previous point: 'a' must have $012_a \vee 034_a$. Therefore, the reasoning of 'c' should conclude at the previous step (4.2.). Nothing else is necessary.

c's final state of knowledge is:

$012_a \vee 034_a$

a = 0; 2   then   b = 5; 4   Therefore 012_a and 543_b
Or
a = 0; 4   then   b = 5; 2   Therefore 034_a and 521_b

As the deal has been: $C_{(a)}{}^3{}_7 * C_{(b)}{}^3{}_4 * C_{(c)}{}^1{}_1 = aR(^3{}_7); bR(^3{}_4); cR(^1{}_1) = 140$ deals are possible. At the beginning, 'c' knows she holds card 6, so the possible deals are only the ones where 'c' holds card 6, therefore there are 20 possible deals: $C_{(a)}{}^3{}_6 * C_{(b)}{}^3{}_3 = aR(^3{}_6); bR(^3{}_3) = 20$. In fact, Cath is only hesitating between two possible deals $[(012_a \wedge 543_b) \vee (034_a \wedge 521_b)]$, thus 'c' knows 18 deals are not possible. If 20 unknown deals mean 'c' is 100% ignorant, then 2 unknown deals mean 'c' is 10% ignorant. In this case, Cath has reached 90% of knowledge, according to the possible deals, and 33.3333% of knowledge of the composition of each possible deal because 'c' knows one card from 'a' (0) and one from 'b' (5) as seen in point 4.2.

Total c's knowledge is 93.3333%   Total c's ignorance is 6.6667%

Exercise 4.73 (A six hand solution): Assume deal of cards 012.345.6. Show that the following is a solution: Anne announces: "I have one of {012, 034, 056, 135, 146, 236}" and Bill announces "Cath has card 6".

Bill (b) opponent		Ann (a) proponent		Cath (c) 'mute guest'
2	!6_c	!($012_a \vee 034_a \vee 056_a \vee 135_a \vee 146_a \vee 236_a$)	1	???

What is 'c' thinking after a's and b's announcements?
1. How did 'b' know I have card 6?
2. If 'b' said 6 and no other out of the four possible cards, those ones 'b' does not have, must be because:
    2.1. In triads announced by 'a' where card 6 is there must be also at least one of b's cards so that 'b' can recognise these triads as *not* a's triads. Therefore, 'a' will never have any binomial accompanying card 6.
    2.2. Besides, each one of the triads mentioned in 2.1. should also guarantee the safety of the announcement, since no player should be inside the 'truth zone' of another one (you can only have total control of your 'truth zone'). To be sure *not* to trespass another player's 'truth zone', the only possible solution, for our case, is to include *only one* real card number of the announcer in each one of these triads.
    2.3. 'c' summarises her reasoning: In the triads where 6 is, there is also a card from 'b' (as seen in 2.1.). Thanks to this, 'b' knows these triads as *not* a's triads. These triads contain also one of a's cards to guarantee the safety of the announcement (as seen in 2.2.).
3. Thus, 'c' removes these triads from a's announcement, and the result is a 'new' a's announcement in c's mind: !($012_a \vee 034_a \vee 135_a$).
4. 'c' asks herself what is a's hand? To answer, she will be doing the following reasoning:

4.1. In the triads where 6 is (056; 146; 236), there is also one card of 'a' and another of 'b', so:

$a = 0; 1; 2$ then $b = 5; 4; 3$
$a = 5; 4; 3$ then $b = 0; 1; 2$
$a = 0; 1; 3$ then $b = 5; 4; 2$
$a = 5; 4; 2$ then $b = 0; 1; 3$
$a = 0; 4; 3$ then $b = 5; 1; 2$
$a = 5; 1; 2$ then $b = 0; 4; 3$
$a = 0; 4; 2$ then $b = 5; 1; 3$
$a = 5; 1; 3$ then $b = 0; 4; 2$

4.2. Next step in c's reasoning is comparing the above triads (4.1.) to the three triads remaining in c's mind (as seen in point 3): $012_a \lor 034_a \lor 135_a$. The result is that every one of them is possible because all are compatible with the condition to include one card from 'a' + one card from 'b' + card 6.

c's final state of knowledge is:

$012_a \lor 034_a \lor 135_a$

$a = 0; 1; 2$ then $b = 5; 4; 3$ Therefore 012_a and 543_b
Or
$a = 0; 3; 4$ then $b = 5; 1; 2$ Therefore 034_a and 512_b
(0; 4; 3)
Or
$a = 1; 3; 5$ then $b = 0; 4; 2$ Therefore 135_a and 042_b
(5; 1; 3)

As the deal has been: $C_{(a)}{}^3{}_7 * C_{(b)}{}^3{}_4 * C_{(c)}{}^1{}_1 = aR({}^3{}_7); bR({}^3{}_4); cR({}^1{}_1) = 140$ deals are possible. At the beginning, 'c' knows she holds card 6, so now the possible deals are only the ones where 'c' holds card 6, therefore there are 20 possible deals: $C_{(a)}{}^3{}_6 * C_{(b)}{}^3{}_3 = aR({}^3{}_6); bR({}^3{}_3) = 20$. In fact, Cath is only hesitating among three possible deals $[(012_a \land 543_b) \lor (034_a \land 125_b) \lor (135_a \land 042_b)]$, thus 'c' knows 17 deals are not possible. If 20 unknown deals mean 'c' is 100% ignorant, then 3 unknown deals mean 'c' is 15% ignorant. In this case, Cath has reached 85% of knowledge. In this instance 'c' cannot reach more knowledge because no card is common to all three of a's possible deals.

Total c's knowledge is 85%   Total c's ignorance is 15%

Exercise 4.74 (A seven hand solution): Assume deal of cards 012.345.6. Show that the following is a solution: Anne announces: "I have one of {012, 034, 056, 135, 146, 236, 245}" and Bill announces "Cath has card 6".

Bill (b) opponent		Ann (a) proponent		Cath (c) 'mute guest'
2	!6$_c$	!(012$_a$V034$_a$V056$_a$V135$_a$V146$_a$V236$_a$V245$_a$)	1	???

What is 'c' thinking after a's and b's announcements?

Now, after doing the previous exercises, 'c' has reached a quite refined method. She knows the procedure is:

1. To take off the triads where her card, 6, is. Then, for this case, the 'new' a's announcement, in c's mind, would be:
   !(012$_a$V034$_a$V135$_a$V245$_a$).
2. She also knows it is not necessary to do step 4.3. After doing it for the first exercise, she considers it useless.
3. Once she knows how the 'new' a's announcement looks (an announcement that does not contain her card in any triad), she needs to compare the resultant possible (a-b)'s deals to the announced triads containing 6, her card.

   (a-b)'s possible pairs, according to a's announcement are:

   a = 0; 1; 2   then   b = 3; 4; 5
   a = 0; 3; 4   then   b = 1; 2; 5
   a = 1; 3; 5   then   b = 0; 2; 4
   a = 2; 4; 5   then   b = 0; 1; 3

After comparing a's possible deals to the triads including 6 (056; 146; 236) no new knowledge is gained, because all of a's possible deals are compatible with the condition for triads with card 6 (one card from 'a' + one card from 'b' + the card from 'c' –card 6). Thus, final c's state of knowledge is:

012$_a$V034$_a$V135V245$_a$

a = 0; 1; 2   then   b = 3; 4; 5   Therefore 012$_a$ and 345$_b$
Or
a = 0; 3; 4   then   b = 1; 2; 5   Therefore 034$_a$ and 125$_b$
Or
a = 1; 3; 5   then   b = 0; 2; 4   Therefore 135$_a$ and 024$_b$
Or
a = 2; 4; 5   then   b = 0; 1; 3   Therefore 245$_a$ and 013$_b$

As the deal has been: $C_{(a)}^3{}_7 * C_{(b)}^3{}_4 * C_{(c)}^1{}_1 = aR(^3{}_7); bR(^3{}_4); cR(^1{}_1)=140$ deals are possible. At the beginning, 'c' knows she holds card 6, so now the only possible deals are the ones where c holds 6, therefore there are 20 possible deals: $C_{(a)}^3{}_6 * C_{(b)}^3{}_3 = aR(^3{}_6); bR(^3{}_3)=20$. However, Cath knows only four deals are possible $[(012_a \wedge 345_b) \vee (034_a \wedge 125_b) \vee (135_a \wedge 024_b) \vee (245_a \wedge 013_b)]$, thus 'c' knows 16 deals are not possible. If 20 unknown deals mean 'c' is 100% ignorant, then 4 unknown deals mean 'c' is 20% ignorant. In this case, Cath has reached 80% of knowledge. In this instance 'c' cannot reach more knowledge because no card is common to all four of a's possible deals.

Total c's knowledge is 80%   Total c's ignorance is 20%

At this point we assume we are ready to deal with the question regarding the 'mute guest', because the rest of the proposed exercises by van Ditmarsch do *not* appear to be relevant any more now that we know the method, so its application will be similar every time. Even increasing the difficulty of the algorithm –that is, increasing equally the number of cards for players 'a' and 'b' while 'c' stays with one (3:3:1; 4:4:1; ...; n:n:1), or in ascending arithmetic progression in which the difference in number of cards between players 'a'-'b' and 'c' is -2 (3:3:1; 4:4:2; ...; n:n:n-2)–, the method would be applied in the same way:
  - For the first case, (3:3:1; 4:4:1; ...; n:n:1), one of the announced arrays must be the proponent's deal and the rest must contain *at least one* and a *maximum* of n-2 elements –cards– from the proponent's deal, to keep always place for a minimum of one and a maximum of two cards from 'b', or one from 'b' and one from 'c'.
  - For the second case, (3:3:1; 4:4:2; ...; n:n:n-2), one of the announced arrays must be the proponent's deal and the rest must contain: *only one* element –card– from the proponent's deal to be sure that at least one card from 'b' is always included in each one of them.

What will change for each particular case is the dimension of the arrays –which must always 1xn, being 'n' the size of the deal for (a-b)– and the number of announced arrays needed to keep the announcement safe.

Thus, the question would be *whether the 'mute guest' is a true passive knowledge subject or a 'hacker in disguise'*. We think the answer is quite clear. As far as the 'mute guest' is really mute but *not* deaf, we cannot be so arrogant as to think that the 'mute guest' is *not* thinking about what they are hearing. We can never assert that they are just hearing and *not* listening. If they are listening, they could be thinking about it. If they are thinking, they will then reach some amount of knowledge; they can be passive (hearing and not listening, then not thinking) or *not*, that is nothing but their choice. So, a 'mute guest' is *not* a real passive knowledge subject because of being mute.

Therefore, the chance to have a 'hacker in disguise' hidden as the 'mute guest' is quite high because: the hacker's performance is just to be hidden; to be 'mute' while others are talking; listening and *not* just hearing; gathering information from the others during the information exchanges; thinking why something is said and *not* something else, *or* in another words. Then, if flaws are found (*flaws* meaning *information leaks*, the information which is said in an unsafe form –including silence), they may decide to start the attack or not. Professional attacks are *not* done at random, they are done with some degree of previous knowledge, and knowledge about others is only obtained from themselves. The hacker's job is *no* other than catching the leaking information and taking advantage of it, using it to conduct a more effective attack. We must be careful, since even when information is passed in a safe form, we cannot be sure that there is no information leaking. Information is information anyway; even silence provides information, because silence encodes usually a finite number of replies to the question that triggered the silence (as already seen in the previous act). Thus, a 'mute guest' is *not* the best guest when you want to pass information without being recorded.

Anyway, the existence of 'mute guests' enhances the argumentative capacities of the negotiator, they are required to do their announcements as correctly as possible (both in quality and safety). Being forced to work under such high standards, a 'mute guest' could be the best coach for a negotiator. In the next chapter we will refer to two relevant lessons learned from the 'mute guest', addressed to improving negotiating capacities.

Chapter 3.2: Two lessons from the 'mute guest'

Being coached as an negotiator in the presence of a 'mute guest', a potential hacker, helps us know what and why are they thinking based on our announced arguments; therefore we can enhance our way of communicating, improve the structure of our announcements –their quality and safety– from the point of view of a syntactic pattern in which the content is expressed. A potential hacker is the best mirror we can have; by observing them we can learn a great deal about potential flaws in the negotiation we are preparing, because a hacker is nothing but the worst opponent.

The two great lessons from the 'mute guest' are the following:
1. The potential hacker starts thinking 'hard' against us when we directly trespass their 'truth zone' (e.g.: Bill announces: "Cath

has card 6"). When somebody feels their 'truth zone' directly trespassed, the natural reaction is to think: *Why do they say that? How can they know?* Everybody's 'truth zone' is the core of their comfort zone, and nobody likes it to be trespassed, and much less so with a direct allusion. When somebody feels overstepped, they feel in danger. Then there are only two possible reactions: either fighting back against this invasion –and the negotiation process is automatically stopped–, or transforming this unintended direct attack into the hardest counterattack we can expect because, as we have seen in the previous chapter, announcing our opponent's truth is the least safe we can do, it is the most revealing action we can make, it shows much more of our 'truth zone' than talking about our proper truth, as Ann did in her announcement.

Thus, this is the first lesson: The 'mute guest' says: "Do not touch me, please, or at least not shamelessly". That is, if you want to remain safe when negotiating, you must be aware that you are *only* speaking the truth, because a lie may be the hacker's truth, and then the hacker will be able to deduce what you want to conceal.

2. Their second lesson is in correspondence with the previous one. Now we know that it is not safe to trespass our opponent's 'truth zone' directly, then *How will we be able to attack and remain safe?* The best plausible way would be to create our replies to the proponent's announcements following a similar pattern to the one that we would use to reply a partner in the presence of a hacker. The question is how Bill could answer Ann and *not* increase Cath's already acquired knowledge (from Ann's announcement). The way to do it is just the one we use naturally when we give information *not* to be understood by a third person: we usually reply repeating the same pattern used before, like *going along with the same* –but it is *not* quite the same–, for instance, Ann announces: "I have one of {012, 034, 056, 135, 146, 236, 245}", then Bill's reply could be just the inverse of the part of announcement already caught by Cath. Thus, Bill's reply could be: "I have one of {543, 125, 056, 042, 146, 236, 013}". This would add nothing to Cath's knowledge:

$a = 0; 1; 2$ then $b = 5; 4; 3$ Therefore 012_a and 543_b
Or
$a = 0; 3; 4$ then $b = 1; 2; 5$ Therefore 034_a and 125_b
Or

a = 1; 3; 5   then   b = 0; 4; 2   Therefore 135_a and 042_b
Or
a = 2; 4; 5   then   b = 0; 1; 3   Therefore 245_a and 013_b

Therefore,

$[K_a(mnp_a) \wedge K_a(m'n'p'_b)] \wedge [K_b(m'n'p'_b) \wedge K_b(mnp_a)] \wedge [K_c(m''_c) \wedge \neg K_c(mnp_a) \wedge \neg K_c(m'n'p'_b)]$

> This is not exactly so; however, Cath has no more knowledge than she acquired after Ann's announcement.

Thus, this is the second lesson: If you want to remain safe from the opponent's attack, when you are continuing with an argument you must create a new piece of information following the same pattern used before.

We cannot be so naïve as to believe that we will not touch the 'truth zone'/comfort zone of our opponent when we are negotiating. We should *not* undervalue our opponent, and we should prepare the negotiation as if it were a struggle against an intelligent hacker; by doing so, we will do the best we possibly can to stay safe.

It is time to check whether some more light has been thrown on the darkness.

Chapter 3.3: A Dialogical Framework. A light in the darkness

As we said in the beginning of this act, our main aim here is to throw some light on the darkness; maybe the dark part could be reduced.
- Direct shadows:
    - *What arguments to choose*. We could definitely say that we cannot be completely sure about what arguments, in the meaning of their content, should be chosen for a negotiation. However, now we know how to choose the best structure (syntactic pattern) for arguments, since we have learned it from the Dialogical Semantics.
    - *To assign to each argument the right card to symbolise it*. Nothing else can be said, as already discussed in page 33.
- Indirect shadow:
    - *Subjectivity of the interlocutor*. We cannot guarantee what will be replied to our arguments. We cannot predict their content. However, in the light of this dialogical framework we can now

predict the most plausible/logical syntactic structure for each reply. Thus, we will be able to know how the agreement could be, depending on the syntactic structure of the proponent's first announcement.

It is definitely better to be the proponent, since they hold the reins of the negotiation dialogue. Nevertheless, this is not always the case: when somebody is coming to see you, good manners usually require asking them first. For this case, different negotiation scenarios have to be prepared because, even if you are able to guess what your opponent will say, you are never sure until they start the discussion. Besides, during a negotiation process the positions –proponent/opponent– are interchangeable, depending on the matter being discussed at the moment.

In any case, it is always better to use the dialogical semantics form that is the most favourable to us. Therefore, when we are acting as proponent, we will be able to predict our defence, and then it will be preferable to choose from structures in green in the table next page, in order to build our arguments without giving weapons to our opponent –structures in red in the table.

As a rule, it can be said that it is better to make announcements under a disjunctive form, a particular form, a possible form, or a combination of them. Moreover, in the specific case of a conditional announcement, it is *not* possible to have it under a completely literal expression: literal antecedent *and* literal consequent, understanding *literal* as an elementary proposition or its negative mood; at least one of them –antecedent or consequent– must be non-literal. In any case, the best choice is to use:

- The consequent under one of these forms we mentioned just above, because then we will receive a 'favourable' attack, since these are the cases where the defender has the choice; and
- The antecedent as an elementary proposition, a conjunctive form, a universal form, a necessary form or a combination of them, because then we will be able to fight back –once our conditional has been attacked, since these are the cases where the attacker has the choice.

Finally, it should always be kept in mind the way in which negative sentences work in our mind (pp. 30-31). We should be extremely careful about this because, in intuitionistic Logic, no defence is possible after a negative sentence has been said and attacked (pp. 38-39).

Dialogical Semantics Form

Announcement structure	Attack	Defence
X !α∧β The attacker chooses the defence	**Y ?L∧** (Left side of the conjunction)	**X !α**
	Y ?R∧ (Right side of the conjunction)	**X !β**
X !α∨β The defender chooses the defence	**Y ?∨**	**X !α**
		X !β
X !α→β	**Y !α** (α is assumed to occur)	**X !β**
X !¬α	**Y !α**	--------- (No defence, *only* counterattack is possible)
X !∀ₓAₓ	**Y ?k** (k is chosen by the attacker)	**X !A_k**
X !∃ₓAₓ	**Y ?∃** (could you show me one, please?)	**X !A_k** (k is chosen by the defender)
X !□A_cl	**Y ?cj** <cIRcj> (cj is chosen by the attacker)	**X !A_cj**
X !◊A_cl	**Y ?◊** (could you show me a case, please?)	**X !A_cj** <cIRcj> (cj is chosen by the defender)

Player **X**: Opponent or Proponent   Player **Y**: Opponent or Proponent

Note: '□' and '◊' will operate in the same way in all the cases where there is a modal operator: alethic, deontic, epistemic, doxastic, temporal or a combination of them.

We have gone deeper in syntax and extracted a pattern to make announcements (attacks or defences). Other than that, there is nothing else in a rational framework, which may throw light on our shadows.

Through this research, we have been able to clarify some dark aspects of the negotiation process. Now, in the next section –Conclusion– we will propose a protocol to deal with negotiations.

CONCLUSION: Negotiating with a Logical-Linguistic Protocol in a Dialogical Framework

In the beginning of this work we said that this was an attempt to think how to create logical dialogues to tackle negotiations, meaning solving conflicts from basic linguistic structures (conjunctions, disjunctions, conditionals) placed under a dialogue form as a cognitive system which 'understands' natural language and where there is a permanent feedback between both.

Now it is time to know whether this can be done and how to do it. This Conclusion will show a possible map, a guide to choose the order of arguments in negotiations with the aim to reach the highest intersection point between the optimization function and the satisfaction function. Since all negotiations start with a decision –the decision of negotiating about something– and as the rationality of the decision remains in the side of objectivity, that is, in optimization, we will offer in this section a sort of protocol that –we consider– could be useful to reach that aim. As a result, satisfaction will increase in a direct proportion and we will be able to reach a high intersection point.

A protocol for negotiating in a dialogical framework

As it is not possible to negotiate in a non-dialogical framework, here we are *not* taking *'dialogical'* in its strict sense as in logical semantics – although that sense is also included–, that is why it is written between quotation marks. We want to remark that *no* negotiation process could exist without dialogue, since dialogue is the *only* form under which a negotiation can take place, can be feasible. Therefore, we will offer here a sort of path to deal with it. For that purpose, we can imagine the different steps a negotiation could go through before sitting at the negotiation table, and how a correct negotiation could be prepared:

1. Getting as much information about the subject (object/s and goal/s) as we are able to obtain.
2. Analysing every piece of information, even the very small ones. All of them are crucial for preparing a right negotiation process. The more we can understand *what, how* and *why* is happening, the better will be the protocol we will design to tackle the negotiation and the better the arguments we will choose to deal with it.

[Point 1 and 2 are the grounds: the better they are done, the more solid the negotiation arenas in these two directions: to reach success and to analyse it afterwards in order to learn for future negotiations. In this work, these points are represented in the Annex, a past case because, as we already said, future cases cannot possibly prove a hypothesis. We needed to test this protocol in a completed case in order to know its potential for future cases].

3. Assessing whether the object/s to negotiate have the same internal structure as the objective/s, that is, the issue/s are in strict correspondence to the goal/s (an illustrative example was given in pages 24-25).
4. Evaluating where is the power, how close to a particular negotiator it is.
5. Knowing whether the scenario is stable enough, whether we have control over the negotiation time and space, or whether they are rather unpredictable.

[If points 3, 4 and 5 are favourable to us –goal/s have the same internal structure as the issue/s to negotiate in order to reach it/them, power is not clearly on one side, and the arenas are solid enough–, then we will be ready to follow the path we are proposing here. If this is *not* the case, then we will *not* be able to continue on this path, as shown in pages 24-25].

6. Taking points 3, 4 and 5 as favourable, it is time to make the whole list of the arguments, including the interlocutor's arguments (of course we cannot know them in advance, but we need to imagine the scenario by making plausible suppositions). At the moment, we are paying attention only to the content of arguments; the structure in which they will be expressed is *not* relevant *yet*. In order to choose the content it will be a great help to know as precisely as possible the beliefs and values of our interlocutor. That will be an invaluable help to build a common territory of beliefs and values in which the arguments will be born (pp. 31-32).
7. Classifying the arguments according to the different aspects shared by the issue/s and the goal/s. In our case –*Camp David Accords*, as an international negotiation– four aspects can be identified: political, financial, social, and cultural (obviously, this has to be adapted for every negotiation area).
8. Placing in order the classed arguments according to their importance, from highest to lowest, in every selected aspect.
9. Translating the arguments (unquantifiable values) into Bridge cards (countable values). So, if we decide, for our case, that spades represent the political aspect, then the most relevant argument will become the Ace of spades, and so on (pp. 15-20).

10. Playing a Bridge hand (pp. 20-22).
    The objection that a negotiator will not be able to use this protocol unless they can play Bridge, is easy to overcome, since there is a computer program that works the opposite way: it chooses the deal and it makes the game. At this moment, the algorithm supporting the program has not been developed to its maximum. However, it should be possible to enhance it because Bridge is just mathematics (cf. Borel & Chéron, 2009), and a Bridge computer program prototype has already been created and is running –you may take a look at the Bridge Base Online (BBO) website: http://www.bridgebase.com/ and see how the robot-player works. Once the algorithm is working one way, it will also be possible that it works in the opposite direction. The reverse option is *not* available, because, for a Bridge player, that will be the antithesis of the game: if you know the result in advance, then what to play for?
11. Analysing what happened during the game (pp. 16-22). This means, to undo the game in order to know how the arguments have been played, discovering the pairs. That will provide a key on what is a win, what is a loss, and what card does *not* have a correlative argument –in case one player holds one suit longer than the other player. For this case, it is better to disregard all the non-correlative arguments when the negotiation takes place, because it is not elegant to ask somebody something they do not have an answer for, and it is also quite impolite to change the subject when being questioned.

[Act I is the main frame for points 3 to 11].

This protocol still needs to be improved. There is excessive subjectivity in the choice of arguments. The quality of the optimization function should be increased while avoiding subjectivity as much as possible, in order to guarantee a good ratio of satisfaction as a result of a well done optimization.

12. Paying attention to the structure –syntax– of the arguments, since negotiating is not throwing one card on top of another on a table. *Negotiation* means dialogue, and so, countless variables. Once we know our possibilities in terms of profit and loss account, we must prepare extremely well each argument that will be used, because we know that the rational truth-value is in syntax (p. 29). Therefore, we will need to apply a subroutine:
    12.1. Analysing the result of every trick, of every pair of arguments, always respecting the order in which they have been played.

12.2. Being aware of our mind presuppositions in front of a conditional sentence. A conditional pattern presupposes always a cause–effect relationship (p. 30) where, if the cause is the case, then the effect will also be (p. 29).
12.3. Taking care *not* to trespass directly the interlocutor's 'truth zone'; therefore, we should always talk from our 'truth zone'. In case some of the used arguments trespass directly the comfort zone of our interlocutor, they need to be modified (pp. 49-51).
12.4. Staying safe. Our arguments must be as concealed as possible, meaning that information *not* strictly needed should *not* be given. The information given should *not* discover our game more than strictly necessary (pp. 50-51).
12.5. Creating the dialogical form for every argument. As we already know the pairs –the tricks–, we know who will act as proponent and who as opponent for each trick. Therefore, we are able to create the right dialogical form for each pair. In any case, it is highly recommended to prepare logical dialogues in both directions, to be ready just in case. At this point we will use *intuitionistic* Logic because, a negotiation being a process, negotiators remain always engaged in the chain of arguments; thus the best way to seek an agreement is *not* saying something now and the opposite later. This is why *intuitionistic* Logic is more commonly used. Dialogues in a classic Logic framework are restricted to the cases in which, after a chain of reasoning, a negotiator thinks that switching the point of view could be better to reach the goal, and the other negotiator agrees.

It is now the time to evaluate the degree of accommodation between the previous research and the path proposed in these pages. It is the time to decide whether it is worthy to carry out into practice such an attempt to improve the negotiation process, tackled from basic linguistic structures (conjunctions, disjunctions, conditionals) placed under a dialogue form as a cognitive system which 'understands' natural language and where there is a permanent feedback between both. Now it seems relevant to go to the Annex, as testing ground, remembering what we said earlier (page 2): "To preserve the rigour and the aseptic nature of this research we have *not* applied any framework susceptible of being applied later (…). Thus, you will not find any application of the *Game Theory*, neither NLP nor Dialogical Semantics along this analysis of *Camp David Accords*". We can now confirm that, out of the three, only Bridge (*Game Theory*) has been susceptible to be properly applicable to the study case. NLP and Dialogical Semantics could have been applied if we had had access to the complete and faithful transcription of all the dialogues in *Camp David* (document *not* available to the public). Throughout the exhaustive study of this case, even with this limitation, we are sure you will be able to make a right assessment on the pertinence of using NLP and Dialogical Semantics for tackling negotiations. Along the study of *Camp David Accords* you will be able to discover what kind (content) of arguments were used, and thus, to suppose how they were expounded with a high degree of certainty. The whole transcription of the dialogues is *not* available but there is a sort negotiation diary. You will find some references to it along this research and its bibliographic whole datum at the Annex Bibliography —Camp David Accords: Thirteen Days After Twenty-Five Years, 2003.

Once we have discovered a new opportunity to apply Logical Dialogues, this time to deal with Negotiations, to solve conflicts and even to serve peace, we would like to point out some possible research lines to continue investigating this particular approach to deal with negotiations. The following lines are open:
- A dialogical analysis for intra-negotiations (inside the team itself), since here we have only considered the inter-negotiations. The author is already thinking on methods to develop this approach, and hopes to start soon.
- A Hintikka's *GTS* (Game-Theoretical Semantics) approach for inter-negotiations and for intra-negotiations as well.
- Computer research: algorithms applied to building dialogues in a fast and easy way into their dialogical form, including the 'hacker's' dialogue. This is probably the most interesting thing to do in order to get ready for possible attacks.
- Philosophical implications of tackling negotiations with logical-linguistic protocols in a dialogical framework.

Bibliography

MANUALS:

Borel, E., & Chéron, A. (2009). *Théorie Mathématique du Bridge à la Portée de Tous*. Paris, France: Éditions Jacques Gabay.

Centre of Excellence (2014a). *NLP Practitioner Course*. Manchester, UK: Centre of Excellence.

Centre of Excellence (2014b). *NLP Study Guide*. Manchester, UK: Centre of Excellence.

Gamut, L. T. F. (1991). *Logic, Language and Meaning* (Vol. 1 et 2). Chicago, USA: The University of Chicago Press.

Kast, R. (1993). *La Théorie de la Décision*. Paris, France: Éditions La Découverte.

Lewicki, R., Barry, B., & Saunders, D. (2004). *Essentials of Negotiation*. New York, USA: McGraw-Hill.

Lorenz, K., & Lorenzen, P. (1978). *Dialogische Logik*. Darmstadt, Germany: Wissenschaftliche Buchgesellschaft.

Mestanza-Fragero, M. (2007). *Bridge. Cultura, Ciencia y Deporte*. Vitoria, España: Galasarguin, S. L.

Marsh, P. D. V. (1984). *Contract Negotiation Handbook*. Hants, UK: Grower Publishing Company Limited.

Peters, H. J. M. (1992). *Axiomatic Bargaining Game Theory*. Dordrecht, The Netherlands: Kluwer Academic Publishers.

United States Conference of Catholic Bishops. (2011). *The New American Bible, Revised Edition (NABRE)*.
   Retrieved from http://www.usccb.org/bible/index.cfm

Van Ditmarsch, H., van der Hoek, W., & Kooi, B. (2008). Dynamic Epistemic Logic. In V. F. Hendricks, J. Symons, J. Hintikka et al. (Eds.), *Synthese Library, 337*. Dordrecht, The Netherlands: Springer.

Wall, J. A. Jr. (1985). *Negotiation: Theory and Practice*. Glenview, Illinois (USA): Scott, Foresman and Company.

White, D. J. (1970). *Decision Theory*. Chicago, USA: Aldine Publishing Company.

Zartman, W., & Berman, M. R. (1982). *The Practical Negotiator*. New Haven, Connecticut (USA): Yale University Press.

BOOKS:

Bicchieri, C. (1993). *Rationality and Coordination*. New York, USA: Cambridge University Press.

Brams, S. J., & Taylor, A. D. (1996). *Fair Division: From cake-cutting to dispute resolution*. Great Britain: Cambridge University Press.

Camerer, C. F. (2003). *Behavioral Game Theory: Experiments in Strategic Interaction*. Princeton, New Jersey (USA): Princeton University Press.

Chamoun-Nicolás, H. (2008). *Negociando como un Fenicio*. Kingwood, Texas (USA): Keynegotiations, LLC

Covey, S. R. (2011). *The 3rd Alternative. Solving Life's Most Difficult Problems*. London, UK: Simon & Schuster.

Fedorov, R., Belov, A., Kovaldzhi, A., & Yashchenko, I. (Eds.). (2011). *Moscow Mathematical Olympiads, 2000- 2005*. Providence, Rhode Island (USA): Mathematical Sciences Research Institute & American Mathematical Society.

Hofstadter, D. R. (1987). *Gödel, Escher. Bach. Un Eterno y Grácil Bucle*. Barcelona, España: Tusquets Editores.

James, B. (2011). *DO IT! or DITCH IT*. London, UK: Ebury Publishing.

Joyce, J. M. (1999). *The Foundations of Casual Decision Theory*. Cambridge, UK: Cambridge University Press.

Morin, E. (1986). *El Método. Tomo I: La naturaleza de la Naturaleza*. Madrid, España: Cátedra.

Petruzzi, J. (2012). *Going for Gold*. Peterborough, UK: FastPrint Publishing.

Rahman, S., & Clerbout, N. (2015). *Las Raíces Dialógicas de la Teoría Constructiva de Tipos*. Retrieved from https://halshs.archives-ouvertes.fr/halshs-01238172/document, checked on August 29, 2019.

Raiffa, H. (1973). *Analyse de la Décision. Introduction aux Choix en Avenir Incertain*. Paris, France: Dunod.

Raiffa, H. (1996). *Lectures on Negotiation Analysis*. Cambridge, Massachusetts (USA): PON Books.

Raiffa, H. (2000). *The Art and Science of Negotiation*. Cambridge, Massachusetts (USA): Harvard University Press.

Redmond, J., & Fontaine, M. (2011). How to Play Dialogues. An Introduction to Dialogical Logic. In S. Rahman (Ed.), *Dialogues and Games of Logic. 1*. London, UK: College Publications.

Van Benthem, J. (2011). *Logical Dynamics of Information and Interactions*. Cambridge, UK: Cambridge University Press.

Wittgenstein, L. (1922). *Tractatus Logico–Philosophicus*. Retrieved from https://www.gutenberg.org/files/5740/5740-pdf.pdf, checked on August 29, 2019.

BOOK CHAPTERS and ARTICLES:

Aumann, R. J. (1987). Game Theory. In J. Eatwell, M. Milgate, & P. Newman (Eds.), *The New Palgrave: A Dictionary of Economics, 2*, (pp. 460-482). London, UK: Macmillan.

Avenhaus, R., & Krieger, T. (2007). Formal Methods for Forecasting Outcomes of Negotiations on Interstate Conflicts. In R. Avenhaus, & I. W. Zartman

(Eds.), *Diplomacy Games. Formal Models and International Negotiations* (pp. 123-148). Berlin, Germany: Springer-Verlag Publishers.

Avenhaus, R., & Zartman, I. W. (2007). Introduction: Formal Models of, in, and for International Negotiations. In R. Avenhaus, & I. W. Zartman (Eds.), *Diplomacy Games. Formal Models and International Negotiations* (pp. 1-22). Berlin, Germany: Springer-Verlag Publishers.

Bacharach, M. (1992). The Acquisition of Common Knowledge. In C. Bicchieri, & M. L. Dalla Chiara (Eds.), *Knowledge, Belief and Strategic Interaction* (pp. 285-315). New York, USA: Cambridge University Press.

Bicchieri, C. (1992). Knowledge–Dependent Games: Backward Induction. In C. Bicchieri, & M. L. Dalla Chiara (Eds.), *Knowledge, Belief and Strategic Interaction* (pp. 327-343). New York, USA: Cambridge University Press.

Bishop, R. L. (1975). Game–Theoretic Analyses of Bargaining. In O. R. Young (Ed.), *Bargaining. Formal Theories of Negotiation* (pp. 85-128). Urbana, Illinois (USA): University of Illinois Press.

Brams, S. J. (1979). Faith versus Rationality in the Bible: Game Theoretic Interpretations of Sacrifice in the Old Testament. In S. J. Brams, A. Schotter, & G. Schwödiauer (Eds.), *Applied Game Theory* (pp. 430-445). Würzburg, Germany: Physica-Verlag.

Cross, J. G. (1991). Economic Perspective. In V. A. Kremenyuk (Ed.), *International Negotiation. Analysis, Approaches, Issues* (pp. 164-179). San Francisco, USA: Jossey-Bass Inc., Publishers.

Druckman, D. (2007). Negotiations Models and Applications. In R. Avenhaus, & I. W. Zartman (Eds.), *Diplomacy Games. Formal Models and International Negotiations* (pp. 83-96). Berlin, Germany: Springer-Verlag Publishers.

Dupont, C., & Faure, G. O. (1991). The Negotiation Process. In V. A. Kremenyuk (Ed.), *International Negotiation. Analysis, Approaches, Issues* (pp. 40-57). San Francisco, USA: Jossey-Bass Inc., Publishers.

Ellsberg, D. (1975). Theory of the Reluctant Dualist. In O. R. Young (Ed.), *Bargaining. Formal Theories of Negotiation* (pp. 38-52). Urbana, Illinois (USA): University of Illinois Press.

Faure, G. O., & Rubin, J. Z. (1993). Lessons for Theory and Research. In G. O. Faure, & J. Z. Rubin (Eds.), *Culture and Negotiation: The Resolution of Water Disputes* (pp. 209-231). Newbury Park, California (USA): SAGE Publications, Inc.

Freymond, J. F. (1991). Historical Approach. In V. A. Kremenyuk (Ed.), *International Negotiation. Analysis, Approaches, Issues* (pp. 121-134). San Francisco, USA: Jossey-Bass Inc., Publishers.

Gärdenfors, P. (1992). The dynamics of Belief Systems: Foundations versus Coherence Theories. In C. Bicchieri, & M. L. Dalla Chiara (Eds.), *Knowledge, Belief and Strategic Interaction* (pp. 377-396). New York, USA: Cambridge University Press.

Gruber, T. R. (1993). Toward Principles for the Design of Ontologies used for Knowledge Sharing. *International Journal Human–Computer Studies, 43*, 907-928. Retrieved from http://tomgruber.org/writing/onto-design.pdf, checked on August 29, 2019.

Gutiérrez-Goncet, R., & Whitelock, D. (2012). Sistemas de creencias en ciencia cognitiva: algunas preguntas a nuestros sistemas de conocimiento. In C. Morano-Rodríguez, J. Campos-Acosta, & M. M. Alcubilla-Martín (Eds.), *Ciencia, Humanismo y Creencia en una Sociedad Plural* (pp. 141-149). Oviedo, España: Ediciones de la Universidad de Oviedo y Fundación Castroverde.

Harsanyi, J. C. (1975a). Approaches to the Bargaining Problem before and after the Game Theory. In O. R. Young (Ed.), *Bargaining. Formal Theories of Negotiation* (pp. 253-266). Urbana, Illinois (USA): University of Illinois Press.

Harsanyi, J. C. (1975b). Bargaining and Conflict Situations in the Light of a New Approach to Game Theory. In O. R. Young (Ed.), *Bargaining. Formal Theories of Negotiation* (pp. 74-84). Urbana, Illinois (USA): University of Illinois Press.

Harsanyi, J. C. (1992). Game Solutions and the Normal Form. In C. Bicchieri, & M. L. Dalla Chiara (Eds.), *Knowledge, Belief and Strategic Interaction* (pp. 355-376). New York, USA: Cambridge University Press.

Hintikka, J., & Sandu, G. (1997). Game–Theoretical Semantics. In J. van Benthem, & A. ter Meulen (Eds.), *Handbook of Logic and Language* (pp. 361-409). Amsterdam, The Nederlands: Elsevier Science B.V.

Kiel, R., & Schader, M. (1994). Using dialog-controlled rule systems in a maintenance module for knowledge bases. In P. L. Hammer (Ed.), *Annals of Operations Research, 52: Decision Theory and Decision Systems* by K. Mosler, & M. Schader (Eds.), (pp. 171-180). Basel, Switzerland: J. C. Baltzer AG, Science Plublishers.

Kremenyuk, V. A. (1993). A Pluralistic Viewpoint. In G. O. Faure, & J. Z. Rubin (Eds.), *Culture and Negotiation: The Resolution of Water Disputes* (pp. 47-54). Newbury Park, California (USA): SAGE Publications, Inc.

Lang, W. (1993). A Professional's View. In G. O. Faure, & J. Z. Rubin (Eds.), *Culture and Negotiation: The Resolution of Water Disputes* (pp. 38-46). Newbury Park, California (USA): SAGE Publications, Inc.

Levine, P., & Ponssard, J. P. (1979). Power and Negotiation. In S. J. Brams, A. Schotter, & G. Schwödiauer (Eds.), *Applied Game Theory* (pp. 13-31). Würzburg, Germany: Physica-Verlag.

Magari, R. (1992). Introduction to Metamoral. In C. Bicchieri, & M. L. Dalla Chiara (Eds.), *Knowledge, Belief and Strategic Interaction* (pp. 257-274). New York, USA: Cambridge University Press.

Martínez-Cazalla, M. D. (2012): La Importancia de las Creencias en las Negociaciones Internacionales. In C. Morano-Rodríguez, J. Campos-Acosta, & M. M. Alcubilla-Martín (Eds.), *Ciencia, Humanismo y Creencia en una Sociedad*

Plural (pp. 407-416). Oviedo, España: Ediciones de la Universidad de Oviedo y Fundación Castroverde.

McClennen, E. F. (1992). Rational Choice in the Context of Ideal Games. In C. Bicchieri, & M. L. Dalla Chiara (Eds.), *Knowledge, Belief and Strategic Interaction* (pp. 47-60). New York, USA: Cambridge University Press.

Montague, R. (1970). Universal grammar. *Thesis, 36* (3), 373-398. doi: https://doi.org/10.1111/j.1755-2567.1970.tb00434.x

Mundici, D. (1992). The Logic of Ulam's Games with Lies. In C. Bicchieri, & M. L. Dalla Chiara (Eds.), *Knowledge, Belief and Strategic Interaction* (pp. 275-284). New York, USA: Cambridge University Press.

Nash, J. F. (1975a). The Bargaining Problem. In O. R. Young (Ed.), *Bargaining. Formal Theories of Negotiation* (pp. 53-60). Urbana, Illinois (USA): University of Illinois Press.

Nash, J. F. (1975b). Two-Person Cooperative Games. In O. R. Young (Ed.), *Bargaining. Formal Theories of Negotiation* (pp. 61-73). Urbana, Illinois (USA): University of Illinois Press.

Powell, D. E. (1991). Legal Perspective. In V. A. Kremenyuk (Ed.), *International Negotiation. Analysis, Approaches, Issues* (pp. 135-147). San Francisco, USA: Jossey-Bass Inc., Publishers.

Pruitt, D. G. (1991). Strategy in Negotiation. In V. A. Kremenyuk (Ed.), *International_Negotiation. Analysis, Approaches, Issues* (pp. 78-89). San Francisco, USA: Jossey-Bass Inc., Publishers.

Puppe, C. (1994). Rational choice based on vague preferences. In P. L. Hammer (Ed.), *Annals of Operations Research, 52: Decision Theory and Decision Systems* by K. Mosler, & M. Schader (Eds.), (pp. 67-81). Basel, Switzerland: J. C. Baltzer AG, Science Plublishers.

Raiffa, H. (1991). Contributions of Applied Systems Analysis to International Negotiation. In V. A. Kremenyuk (Eds.), *International Negotiation. Analysis, Approaches, Issues* (pp. 5-21). San Francisco, USA: Jossey-Bass Inc., Publishers.

Rapoport, A., & Kahan, J. P. (1979). Standards of Fairness in 4-Person Monopolistic Cooperative Games. In S. J. Brams, A. Schotter, & G. Schwödiauer (Eds.), *Applied Game Theory* (pp. 74-95). Würzburg, Germany: Physica-Verlag.

Reny, P. J. (1992). Common knowledge and Games with Perfect Information. In C. Bicchieri, & M. L. Dalla Chiara (Eds.), *Knowledge, Belief and Strategic Interaction* (pp. 345-353). New York, USA: Cambridge University Press.

Rubin, J. Z. (1991). The Actors in Negotiation. In V. A. Kremenyuk (Ed.), *International Negotiation. Analysis, Approaches, Issues* (pp. 90-99). San Francisco, USA: Jossey-Bass Inc., Publishers.

Rudnianski, M., & Bestougeff, H. (2007). Bridging Games and Diplomacy. In R. Avenhaus, & I. W. Zartman (Eds.), *Diplomacy Games. Formal Models and*

International Negotiations (pp. 149-179). Berlin, Germany: Springer-Verlag Publishers.

Salacuse, J. W. (1993). Implications for Practitioners. In G. O. Faure, & J. Z. Rubin (Eds.), *Culture and Negotiation: The Resolution of Water Disputes* (pp. 199-208). Newbury Park, California (USA): SAGE Publications, Inc.

Schneider, L., & Cunningham, J. (2003). Ontological Foundations of Natural Language. Communication in Multi-agent Systems. *Lecture Notes in Computer Science, 2773*, 1403-1410.

Schüssler, R. (2007). Adjusted Winner (AW) Analyses of the 1978 Camp David Accords—Valuable Tools for Negotiators? In R. Avenhaus, & I. W. Zartman (Eds.), *Diplomacy Games. Formal Models and International Negotiations* (pp. 283-296). Berlin, Springer-Verlag Publishers.

Sebenius, J. K. (1991). Negotiation Analysis. In V. A. Kremenyuk (Ed.), *International Negotiation. Analysis, Approaches, Issues* (pp. 203-215). San Francisco, USA: Jossey-Bass Inc., Publishers.

Sergeev, V. M. (1991). Metaphors for Understanding International Negotiation. In V. A. Kremenyuk (Ed.), *International Negotiation. Analysis, Approaches, Issues* (pp. 58-64). San Francisco, USA: Jossey-Bass Inc., Publishers.

Shin, H. S. (1992). Counterfactuals and a Theory of Equilibrium in Games. In C. Bicchieri, & M. L. Dalla Chiara (Eds.), *Knowledge, Belief and Strategic Interaction* (pp. 397-413). New York, USA: Cambridge University Press.

Siebe, W. (1991). Game Theory. In V. A. Kremenyuk (Ed.), *International Negotiation. Analysis, Approaches, Issues* (pp. 180-202). San Francisco, USA: Jossey-Bass Inc., Publishers.

Wierzbicki, A. P. (2007). Rationality of Choice versus Rationality of Knowledge. In R. Avenhaus, & I. W. Zartman (Eds.), *Diplomacy Games. Formal Models and International Negotiations* (pp. 69-82). Berlin, Germany: Springer-Verlag Publishers.

Young, O. R. (1975). Strategic Interaction and Bargaining. In O. R. Young (Ed.), *Bargaining. Formal Theories of Negotiation* (pp. 3-19). Urbana, Illinois (USA): University of Illinois Press.

Young, H. P. (1979). Exploitable Surplus in N-Person Games. In S. J. Brams, A. Schotter, & G. Schwödiauer (Eds.), *Applied Game Theory* (pp. 32-38). Würzburg, Germany: Physica-Verlag.

Zartman, I. W. (1991a). Regional Conflict Resolution. In V. A. Kremenyuk (Ed.), *International Negotiation. Analysis, Approaches, Issues* (pp. 302-314). San Francisco, USA: Jossey-Bass Inc., Publishers.

Zartman, I. W. (1991b). The Structure of Negotiation. In V. A. Kremenyuk (Ed.), *International Negotiation. Analysis, Approaches Issues* (pp. 65-77). San Francisco, USA: Jossey-Bass Inc., Publishers.

Zartman, I. W. (1993). A Skeptic's View. In G. O. Faure, & J. Z. Rubin (Eds.), *Culture and Negotiation: The Resolution of Water Disputes* (pp. 17-21). Newbury Park, California (USA): SAGE Publications, Inc.

Zartman, I. W., & Avenhaus, R. (2007). Conclusion: Lessons for Theory and Practice. In R. Avenhaus, & I. W. Zartman (Eds.), *Diplomacy Games. Formal Models and International Negotiations* (pp. 323-338). Berlin, Germany: Springer-Verlag Publishers.

Zartman, I. W. (2009). Conflict Resolution and Negotiation. In J. Bercowitch, V. A. Kremenyuk, & I. W. Zartman (Eds.), *The SAGE Handbook of Conflict Resolution* (pp. 322-339). London, UK: SAGE Publications Ltd.

THESIS:

Magnier, S. (2013). *Considérations dialogiques autour de la dynamique épistémique et de la notion de condition dans le droit* (PhD thesis). University of Lille-3, Villeneuve d'Ascq, France.

Martínez-Cazalla, M. D. (2011). *The Bridge. A bridge toward Negotiations.* (Master's thesis). Catholic University of Louvain (UCL), Louvain-la-Neuve, Belgium.

TALKS:

Van Aerde, M. (March 10, 2016), *Veritas*. Saint Albert Library, Brussels, Belgium.

WEBSITES:

http://www.apastyle.org/ & http://normasapa.net/2017-edicion-6/ Authorized websites for American Psychological Association (APA) style.

http://www.bridgebase.com Bridge Base Online (BBO) –website to play Bridge in real time.

http://www.britannica.com Authorized website of the *Encyclopædia Britannica*.

http://stanford.edu Authorized website of the *Stanford Encyclopedia of Philosophy*.

ANNEX
Study Case: *Camp David Accords*

Table of contents

Preface **73**
INTRODUCTION **75**
Key Pieces (OCEAN) **77**
    Object of the negotiation 77
    Context or the negotiation 79
    Elements of the negotiation 80
    Asymmetrical power among the teams 80
    Negotiators' profile 81
The Threads that weave together the Key Pieces **83**
    Structure of the negotiation 84
    The matter of cultural differences 85
    Relationships among the negotiators 86
    The strategy 87
    The process 88
The Key Piece: The Agreement. Level of Adjustment **93**
CONCLUSION: Logic in the Past, a Lesson for the Future **99**
Bibliography **101**

Preface

Our conclusion about how to establish a protocol to find out the best order to use arguments during a negotiation process had to be proven. We needed a testing ground in which they could be verified. That is why we chose a completed negotiation case to guarantee an objective application, because there is no possibility to alter the events. The case chosen to apply the previous research has been *Camp David Accords*. As we said in the Introduction of Act I of the main work, the research about *Camp David Accords* already appeared in a previous work (*The Bridge. A bridge toward Negotiations*) as Chapter 5 (Martínez-Cazalla, 2011, pp. 38-68)[*], being this research older than its transcription in that chapter. Here we will recover that chapter, but *not* literally. Therefore, the use of square brackets to show additions or alterations to the original text does *not* apply here.

The document in this annex should help assess this research in terms of right or wrong; without this document the semantic truth would be unknown. To remain faithful and preserve the rigour and the aseptic nature of this piece of research we do not apply here any of the frameworks that we analysed earlier, so you will not find any application of the *Game Theory*, neither NLP nor Dialogical Semantics along this analysis of *Camp David Accords*.

*Why were these the agreements reached and **not** others?* As this is the only fact we really know, this is what the present negotiation analysis investigates. To answer this question it is necessary to analyse every piece of information, even the smallest ones; all of them are crucial for the negotiation analysis. When we can understand what happened and why, then, and only then, can we be sure we stand on solid territory to test our conclusion. Therefore, this document is a proper research work by itself. This is why this Annex contains its own bibliography section, where you can find specific sources as well as others already referred to in the bibliography section of the main work; the latter are shared sources for both research approaches.

This part of the work was done before the main piece of research, even before the first work on the subject (Martínez-Cazalla, 2011). Sometime later, when investigating about how to solve the difficulty to decide the order in which to use arguments during a negotiation process, we ratified this case

---

[*] A published presentation by Martínez-Cazalla (2012) about this subject is also available.

was a very good testing ground for checking our results. This is why this document is laid out in a slightly different style, as you will realise in the bibliography section.

It was on that occasion that the LORD *made a covenant with Abraham, saying:*
To your descendants I give this land,
from the Wadi of Egypt to the Great River [Euphrates]
Genesis 15:18[1]

INTRODUCTION

This work has been written thinking of people who love logical[2] games, because talking about a negotiation is keeping in the scene the different pieces involved in the study case and finding the threads that weave them together, then understanding what happened and why.

An investigation which tries to analyse a negotiation must give an answer to the question *Why were the agreements these, and **not** others?* For our specific case, *Camp David Accords*, the concrete question is *Why did the Sinai come back to Egypt and **not** any other of the territories occupied during the '67 War?* This was the fact that made possible the signature of *The Egyptian–Israeli Peace Treaty*, six months later (Washington, March 26, 1979). Knowing and understanding this event is the only possibility to understand the history. Once history is understood, then it is possible to discuss it. When we are discussing history we are making History. The aim of analysing an international negotiation is to learn the lessons that History is teaching. For this purpose, we need to redo the puzzle of the analysed case from the only part that is known: the final agreement –the *Accords*–.

In the following pages we will try to revive the negotiation held during thirteen days in Camp David in September 1978, between Israel and Egypt, with the mediation of the USA. Thus, we suggest you to forget the known agreements –*Accords*– and to start looking for the pieces involved.

Let us start this work with the key pieces (OCEAN). First of all we will have to answer these questions:
- Which was the **O**bject of the negotiation?
- Which was the **C**ontext of the negotiation?
- Which were the **E**lements of the negotiation?

---

[1] United States Conference of Catholic Bishops (1970). Square brackets in the consulted text.
[2] It should be noted that the word *logical* is taken in a wide sense along this piece of research –Annex–, and *not* in the meaning it has in the context of formal Logic.

- Was there an **A**symmetrical relationship of power among the different teams involved?
- Who were the **N**egotiators –including the mediator?

Second, we have to relate these pieces to each other. Our game is started, we have the pieces and now we begin assembling them. We necessarily wonder about the threads which weave together the **O**bject with the **C**ontext with the **E**lements with the **A**symmetrical relationship of power and with the **N**egotiators, seeking an answer the following questions:
- Which was the structure of the negotiation?
- Were there cultural differences among the negotiators?
- How were the relationships among the different negotiators involved?
- Which was the strategy deployed?
- What was the process?

Once the different pieces and their relationships have been addressed, our puzzle seems finished but, is it really finished? This is the moment to remember the *Accords* forgotten in the former page and try to fit them in our puzzle. The tighter the pieces fit, the better the master lesson learned, which will be used to continue building History along future negotiations.

To deal with the questions arisen before, we would normally use specific sources, that is, the official documents of *Camp David Accords*, and the manuals about it or about international negotiations; the huge amount of literature available –not all of it rigorous– makes it impossible to handle within the scope of this paper. Giving our puzzle a logical form required narrowing down the documents from ordinary press sources, as they are usually not specialized. Since the negotiations were held in the most complete privacy, this work has been developed starting from the declassified sources, authorized sources and specialized manuals. As in every choice, there are advantages –working with the most accurate sources–, but also disadvantages –omitting the opinions of the world about what happened in *Camp David–. What is more real, what the world thinks and believes or what the documents say?* People would probably say general opinion is more real; however, we think that working from the direct sources is better to learn how the negotiation was handled. Another disadvantage of working from the original documents is that not all declassified sources are easily available, for instance, the *Briefings to prepare the Camp David Accords*.[3] Another problem was to verify the conclusion to our hypothesis with experts (*Why did the Sinai come back to Egypt and **not** any other of the territories occupied during the '67 War?*), since authorized and declassified sources explain what happened, but *not* why –or not so clearly. To discover the true reasons for the final agreement requires us to be logical, to look thoroughly into every piece of information and to verify the findings.

We invite you to look in the next pages at the different pieces of the negotiation proposed, and to build your personal puzzle with them. Our puzzle and our conclusion, although well founded, are not the only possible truth. This work aims to be just a modest contribution to the search of key pieces that help analyse negotiations.

Please enjoy the paper and remember that the only possible glue for the puzzle pieces is logic[4]. Good luck in this logical game!

Key Pieces (OCEAN)

In this chapter we analyse the key pieces of our puzzle. They are the ones proposed at the introduction of this paper: the **O**bject, the **C**ontext, the **E**lements, the **A**symmetrical power, and the **N**egotiators involved (including the mediator) for this study case *Camp David Accords*. The order proposed to follow is *not* an order of importance; it is *only* an order to put forward the pieces involved, as a framework for our work: building the puzzle of *Camp David Accords*.

Object of the negotiation

It has to be stated first that the object of a negotiation is *not* the objective/goal of the negotiation. The goal is always to find an agreement and the object is the matter that we want (more or less willingly) to negotiate.

---

[3] Vance (1978). *Study Papers for the Camp David Talks* –official name of this document– or *Briefings to prepare the Camp David Accords* –name given by the Carter Library to the collection of documents used by Carter to prepare the *Camp David* meeting with Israel and Egypt in September 1978, when this piece of research was written. These documents were declassified on August 13, 2003 but in the time this piece was written, they were *only* available for consultation at the library room by individual previous request; permission to photocopy had to be also requested at that time. Nowadays, this document is fully available to the public, even online.
[4] It should be noted that the word *logic* is taken in a wide sense along this piece of research – Annex–, and *not* in the meaning it has in the context of formal Logic.

The object to negotiate is the sovereignty of the territories occupied by Israel during the '67 War, because this is the main reason for the lack of peace in the Middle East. Peace between Israel and the Arab Countries, and more specifically peace between Israel and Egypt, is the true objective, because after the '67 War real peace never came, the conflict became permanent in a higher or lower degree, and the situation in the Middle East was more unstable every day. The world needs and wants peace in this region, and the region needs it too. Eleven years after the '67 War the Middle East and the Occidental World wished to set the UN SC RES 242 and 338, each party according only to their own interests.

Let us see now where the key points of this object are; therefore we will look at the essential paragraphs in those two Resolutions of the UN Security Council, which involve controversial points:

From the UN SC RES 242 (November 22, 1967):

> 1. Affirms that the fulfilment of Charter principles requires the establishment of a just lasting peace in the Middle East which should include the application of both the following principles:
> (i) Withdrawal of Israel armed forces *from territories occupied in the recent conflict* [emphasis added];
> (ii) Termination of all claims or states of belligerency and respect for and acknowledgement of the sovereignty, territorial integrity and *political independence of every State in the area* and their right to live in peace within secure and recognized boundaries free from threats or acts of force [emphasis added];
> 2. Affirms further the necessity
> (a) For guaranteeing freedom of navigation through international waterways in the area;
> (b) For achieving a just settlement of the refugee problem;
> (c) For guaranteeing the territorial inviolability and the political independence of every State in the area, through measures including the establishment of demilitarized zones;

From the UN SC RES 338 (October 22, 1973):

> 2. *Calls upon the parties concerned to start immediately after the cease-fire the implementation of Security Council resolution 242 (1967) in all of its parts* [emphasis added],

3. Decides that, immediately and concurrently with cease-fire, *negotiations shall start between the parties concerned under appropriate auspices aimed at establishing a just and durable peace in the Middle East* [emphasis added].

Israel and Egypt sat to negotiate the sovereignty over the territories occupied by Israel during the *'67 War* to consolidate the peace in the Middle East, but the controversial point is there since the first sentences: "(...) from territories occupied in the recent conflict" (UN SC, 1967, RES 242, 1.(i)). Which are the *territories occupied?* Does this mean all territories occupied? or *Will this sentence be fulfilled if the resolution is applied to just one territory?*

For the moment, we can agree that the Object is negotiating the sovereignty of the territories occupied by Israel during the *'67 War*, without going into discussing what the true meaning of "territories occupied" is.

Context of the negotiation

The context of the negotiation is a regional context. The region is the Middle East, a region of paramount importance in the world because of its position: capital geographic position (gravity centre between East and West), capital geopolitical position (directly deriving from the former), capital geological position (because of the riches in its subsoil), and capital geoeconomical position (a consequence of the geological wealth). Anything that happens there will be transcendent to the rest of the world. Therefore, the context is at the same time regional and global.

It should be kept in mind that the negotiation occurs in the context of the *Cold War era*. The support given by the USSR to the Arab States directly involved in this conflict has been evident since the *'67 War*, while the USA gave support to Israel.

The tension grew higher after the last war, *Yom Kippur War* (October 1973), and the risk of a new world war increased. The conflict won a true international dimension. The decision of the USA to mediate in favour of an accord among the parties involved, specifically between the most powerful parties on the region (Israel and Egypt), was a strategic and diplomatic decision. On one hand, the USA made an attempt to pacify the region, being peace the best guarantee to assure control of the area and its economic resources; on the other hand, as the USA volunteered as mediator, it appeared in front of the world as the saviour and *not* as a threat to its rival, the USSR. Moreover, the USA appeared in front of the parties (Israel and Egypt) as a potential true mediator, in spite of its interests and preferences, and also as the true power in the world —in the '70s the USSR power was already in decadence.

Elements of the negotiation

The elements —interests— of the negotiation are *not* always real, but their importance is that they are true inside the mind which believes in them. What are the elements inside the minds of the negotiators, including the mediator? Answering this question is almost impossible because nobody can be into anybody's mind. In this sense, any answer will be a little daring. Nevertheless, we will try to give a possible answer, based on the background of knowledge that we have about the conflict being negotiated.

In such a way, we can think that the elements involved are:
- *To Israel*: Preserving the territories occupied during the '67 *War*, and consolidating its presence and power in the world. The territories occupied are at the same time a physical item and a psychological item, with all elements that the psychological items entail (in terms of power and cultural presence).
- *To Egypt*: Recovering the territories occupied during the '67 *War*, and so, raising its power in the Middle East, physical power, since it would be the *only* state that could win something from Israel, so much as recovering the control on the two banks of the Suez Canal.
- *To the USA*: To them the element *territories occupied* is a very good pretext, the true element to get involved in the negotiation being their need to ensure peace in the region. The USA has great economic, strategical, and political interests there: on the one hand, to control the power of the USSR, and on the other hand to secure the supply of raw materials to cover its needs.

Asymmetrical power among the teams

Who the mediator was, determined the degree of power asymmetry in this negotiation.

In the context of a bipolar world: *Cold War era —USA vs. USSR*, in which each of the parties (Israel and Egypt) was supported by one of the power blocks (USA or USSR), the power balance fell toward one side from the beginning, even if it was only a psychological effect, due to the fact that the USA was the third negotiating party (the mediator). Moreover, playing at home is always an advantage, and this negotiation took place in the USA. So, it could be said that Israel starts off as a favourite, since the mediator happens to be Israel's own protector. However, the USA mediation will aim to protect first the USA own interests in the Middle East. It is the USA which had *de facto* the power balance in its hands. No wonder they volunteered as mediators, so much for the need to guarantee their interests as for the ability to show-off muscle —we are at the end of the '70s, still in the *Cold War era* period, and the power of the USSR starts showing the large internal fissures that took the USSR to crumbling down a decade later (1989).

Negotiators' profile

Negotiators are a key piece; they are the subjects of negotiation. Therefore, they hold the negotiation together. Until now, we have looked at the invariable keys, which are independent of the negotiator dealing with them. However, their value changes according to the subject who negotiates with them, thus they might increase or diminish their worth. Negotiators are the active part in negotiation, and they will provide gains or losses in the course of negotiation.

Let us review the main features of the profile of each one of the negotiators, in order to understand how our pieces hold together. Those features can be found in the biographical data that we know of each one of the subjects:

- *Menachem Begin for Israel*: Polish origin (Brest, August 16, 1913). Attached to the Zionist movement in Poland since age 12 years. He obtained a Law Degree in 1935 at the Warsaw University. Outstanding student because of his rhetoric and oratory abilities. After the Nazis invasion of Poland he was captured by the Soviets in his flee and was deported to a labour camp in Siberia. After liberation, having lost his family during Holocaust, he moved to Palestine while serving in the British army as an interpreter. He took advantage of this position to actively collaborate in overthrowing British power from Palestine and in the illegal immigration of Jews. After proclamation of the State of Israel he remained very active in the Likud bloc (Jewish right party) reaching leadership in 1977. He was the first Israeli Prime Minister to name the Jordan River territories, including the West Bank, using their biblical names (Judea and Samaria). He was also the first Israeli Prime Minister to set foot on Egypt (Ismailia, December 25, 1977), in correspondence to the visit of Sadat, President of Egypt, to Jerusalem one month earlier —November 1977.

From the former data, we can infer that Begin's personality is of the type "Extraversion—being sociable, assertive, talkative" (Barrick & Mount, 1991. Quoted by Lewicki, Barry, & Saunders, 2004, p. 429).

- *Anwar al-Sadat for Egypt*: Born in Egypt (December 25, 1918), he went to a British military school in Egypt; his aim, however, was to overthrow British power from his homeland. After graduating, he went in search of Nasser, and together they started gathering a revolutionary group of military officials. After being in prison for revolutionary activities and for actively collaborating in King Farouk's deposition –a puppet of the British power–, he finally reached power in the neighbourhood of Nasser –eventually becoming his confidence man. He was vice president since 1969 until Nasser's death in 1970; Sadat took then office as President of Egypt. After their defeat in the *Yom Kippur War* in 1973 and the urgent state of necessity of the country, Sadat reoriented his Foreign Policy. Addressing the force toward the Arab League and the fight against the State of Israel seemed unwise under the circumstances: their backing power, the USSR was showing increasing signs of decadence; Egypt was going through a period of necessity after the two defeats in front of Israel. This situation required accepting the USA help and starting to look toward the West in a more open manner, fighting out suspicion. Sadat, invited by the Israeli Prime Minister Begin, visited Jerusalem in November 1977 with the aim of starting a peace process in the Middle East.

From the former data we can infer that Sadat's personality is of the type "Conscientiousness—being responsible, organized, achievement oriented" (Barrick & Mount, 1991. Quoted by Lewicki et al., 2004, p. 429).

- *Jimmy Carter for the USA* (mediator): Born and raised in the State of Georgia (October 1, 1924), in the South East of the USA. The son of farmers, and active Christian Baptist throughout his life. He holds a degree in Science and was a member of the Office of Naval Intelligence, ascribed to submarine missions. He started his political career in the late '50s. He became President of the USA for the Democrat party on January 20, 1977. Among the priorities for his term was the creation of the Department of Energy –August 4, 1977. He signed a new anti-ballistic accord with the USSR (*SALT II*. Vienna 1979). He also established diplomatic relations between the USA and the Popular Republic of China.

From the former data we can infer that Carter's personality is of the type "Agreeableness—being flexible, cooperative, trusting" (Barrick & Mount, 1991. Quoted by Lewicki et al., 2004, p. 429).

Once the key pieces have been analysed, and once we know the personalities of the negotiators in the play, we may start thinking that the negotiation took place within the tension between *competing* and *collaborating*, with a clear trend toward *competing*. Carter's tough task throughout the negotiations was to conduct such tension toward a result of *compromise*. In fact, Carter was aware of the strong need for *compromise* in the *Briefings to prepare the Camp David Accords*:

> II. <u>How to Mediate Between the two Leaders</u> [underlined in the original text]
> -- Both master manipulators, utilizing basically two different personality styles in order to achieve power and control. Begin concentrates on tactics and details, whereas Sadat focuses on the grand strategy, often employing broad dramatic gestures. In each case, this allows them to avoid making hard decisions. The intermediary trying to bring Sadat the conceptualist and Begin the Talmudic scholar together will have to move each man away from his preferred political (and psychological) style.
> -- In dealing with Begin, avoid entering into word definitions. Allow him to make his basic point without interference and then point him to the intended objective. Begin's concentration on detail is basically an evasive, controlling technique which can be overcome by summarizing succinctly his point of view and then redirecting him back to the mainstream of the discussions.
> In contrast, Sadat will need more guidance, direction, and limit-setting. Left alone, he may get involved in ambiguities and generalities. The President can take advantage of this style by summarizing Sadat's basic intent in such a way that it appears that there are greater points of agreement with Begin than would otherwise be the case. (Vance, 1978 –declassified on August 13, 2003–, TAB 4 –Considerations for Conducting the Summit Meetings–, p. 3)

Now we have the puzzle pieces, we can foresee the degree of *assertiveness↔competitivineness* in which the relationships among the pieces will proceed. This is the time to go one step forward: toward a more thorough analysis of the relationships among those pieces.

The Threads that weave together the Key Pieces

The pieces of this negotiation, impersonated mainly by the two characters, Begin and Sadat, need to reach BATNA (**B**est **A**lternative **T**o a **N**egotiated **A**greement) in order to assemble seamless; however, in sight of the two personalities, the initial point was *bargaining* situation. Changing it into a *win-win* situation would require a long development. Let us now analyse the different aspects that compose the path to BATNA.

As happened with the order of introduction of the key pieces, the aspects of the negotiation introduced hereafter will *not* be presented in order of importance, because each one of them is relevant for the assembling of our puzzle.

Structure of the negotiation

Analysing the structure of a negotiation gives us, as Zartman (1991) puts it, "a skeleton key" (p. 65) in order to understand how the different pieces moved and reached their final relation; we could say that the structure is the scaffold where we can walk, while always keeping in mind the flexibility limits of such a structure.

For our case, we can imagine a triangular scaffold, each one of the actors –Negotiators– in a vertex; Begin and Sadat are in the base vertices, whereas Carter is in the apex. This triangle is circumscribed by the circle of Context. The triangle sides represent the Elements –interests– for which the negotiation takes place. The angular tension is given by the degree of power Asymmetry; in this case there is some tilt in favour of Israel and the USA, as we have seen before, creating two angles <60°, whereas Egypt angle is >60°. The area inscribed in the triangle represents the Object of negotiation.

We have already described our scaffold –skeleton. For the moment, this is an inert shape, it is what is given. How to go from static to dynamic, from the *bargaining* attitude to a *win-win* attitude, will be the task of the mediator. Carter had to find the 'integral function' of the vectors that are present in the triangle, concentrate them in the triangle incenter, and thus achieve the Objective of the negotiation.

We are in front of a circumscribed triangle shape that will be forced in an attempt of changing it into a point circumscribed by a context that will have been modified by the direct action of the integration of the triangle in the point. The change from one scaffold shape to the other will be determined by these four questions: the underlying cultures, the relationships among the actors, the unfolded strategy, and the process. The strategy will be determined by the relationships among the actors, and it will undoubtedly be influenced by the deepest cultural roots of each of them (human beings make decisions based on their beliefs –*not only* religious).

Now it is time we inquire about the cultural background of each negotiator, so to understand later their mutual relationships, and further the strategy that each of them will display during the process. Finally, the analysis of the process will provide us a global vision of what happened and why. We will be able to obtain the 'formula' (the 'integral function' defining the convergence area –incenter–), although with incomplete precision (because that is how integrals are, where the differential coefficient has to be ideally as close as possible to 0); anyway, our 'function' is nothing but a sort of mathematical representation of BATNA.

The matter of cultural differences

The negotiators come from different cultures: Begin from Jewish, Sadat from Muslim and Carter from Christian culture. However, the three of them share something: they are compromised with the religion of their culture. This is their common point; therefore, here lays the basis of the 'function' we want to start building.

We should be aware that the load of the cultural values is stronger in Begin and Sadat than in Carter, because in the Jewish and Muslim cultures there is only a narrow separation between religious and political power. The fact that Carter shares with the others the worth of religion makes him a suitable mediator, since he knows the keys with which religion can re-ligate people. To find a meeting point among these apparently dissimilar cultures was Carter's heavy task:

> At points of resistance, the President may remind them that they already share objectives in common. The Summit meetings are a means of discovering those points of similarity. The objective is to <u>minimize</u> their <u>real</u> differences and <u>maximize</u> their apparent similarities [underlined in the original text]. (Vance, 1978 –declassified on August 13, 2003–, TAB 4 – Considerations for Conducting the Summit Meetings–, p. 1)

The point that would *minimize their real differences and maximize their apparent similarities* is to be found in their deepest, in the main root of their beliefs. Only by knowing this point was it possible to design an adequate strategy for this case, in which the object was the lands and the boundaries between states with different cultures:

> (...) This is an especially challenging (negotiation) because cross-cultural and international negotiations add a level of complexity significantly greater than within-culture negotiations.
> (...) This complexity is a source of frustration for many cross-cultural negotiators, who would like clearer practical guidance when negotiating across borders. (Lewicki et al., 2004, p. 443)

Zartman (1993) states that:

> Culture is indeed relevant to the understanding of the negotiation process —every bit as relevant as breakfast and to much the same extent. Like the particular type of breakfast the negotiators ate, culture is cited primarily for its negative effects. Yet even the best understanding of any such effect is tautological, its measure vague, and its role in the process, basically epiphenomenal. (p. 17)

Moreover, the New Encyclopædia Britannica defines *culture* as:

> (…) the integrate pattern of human knowledge, belief and behaviour. Culture thus defined consists of language, ideas, beliefs, customs, taboos, codes, institutions, tools, techniques, works of art, rituals, ceremonies and other related components; and the development of culture depends upon man's capacity to learn and to transmit knowledge to succeeding generations. (1990, Vol. 3, p. 784. Quoted by Lang, 1993, p. 38)

In the present case, culture (religion, in a more generic sense) is what provides a "(…) bridge between the two sides—" (Salacuse, 1993, p. 204). Thus, this will be our starting point —the experience of the religious fact—; it will be represented by the symbol of the 'integral function' and it will be specifically characterising our 'function'.

Now it is time to consider the relationships of the actors among themselves and with their acts —the strategy deployed which creates a process. We will look at this as 'properties' which are characterising the negotiators' relationships and their actions ('reflexive', 'symmetric' and 'transitive'), seeing them as items constituting one ensemble.

Relationships among the negotiators

Thinking about this brings up directly the question of the behaviour that the negotiators will show and their interactions. The question is really about the 'reflexive property', because the isolation imposed on the negotiators has them in a closed system in which their only relationship is with themselves. Their personalities, strongly characterised by their beliefs in a wide sense, define their relationships.

Thus, we can expect a relationship revolving around itself, marked by the reiterative rhythm of religious fundamentalism of both Begin and Sadat. Their relationship will have a marked authoritarian character, accompanied by the peal of unmatched notes. Carter, understanding this fundamentalism, 'will have to ring the bell', aiming at cracking the dispute in order to open a gate to *integrative* dialogue, breaking the *distributive* rhythm that the others intend to set. With this aim in mind Carter sets the pace of the meetings reaffirming whatever links the negotiators and eluding or postponing whatever divides them:

> The pivotal issue in the talks will be Israel's need to know whether they can get an agreement on the Sinai and what price they must pay for it in concessions on the West Bank. You will want to clarify with Sadat in your opening session that the prospect of an agreement there will be the major incentive for Israel. The more precise he can be about a final, if phased, agreement there the more he can seek in the West Bank/Gaza in return. (Vance, 1978 –declassified on August 13, 2003–, TAB 1 –An Overview of the Camp David Talks–, p. 2)

The strategy

Once we know the nature (culture/religion) that determines the 'reflexive' relationship, we want to know how else this specific nature could express beyond this sterile relationship. That is, we want to know about the strategies on which the actors based their actions. Every cause inevitably produces an effect. Thus, to analyse the strategy developed in a negotiation means to analyse the 'symmetric property' (the actors' relationships and their influence on their actions).

The strategy is just the negotiators' estimation about how they will proceed. It is determined by the personality of each of them and by the goals they wish to reach as result of the negotiation process. So, the strategy will be more or less effective depending on the self-control the negotiators can exert when expressing their personality, and on the definition of their goals:

> Effective goals must be concrete, specific, and measurable. The less concrete and measurable our goals are, the harder it is to (a) communicate to the other party what we want, (b) understand what the other party wants, and (c) determine whether an offer on the table satisfies our goals. (Lewicki et al., 2004, p. 109)

The strategy is the frame in which the negotiation process will develop. The choice of certain tactics to implement the planned strategy will also have an effect in the outcome of the negotiation:

> Although the line between strategy and tactics may seem fuzzy, one major difference is that of scale, or immediacy (Quinn, 1991). Tactics are short-

term, adaptive moves designed to enact or pursue broad (or higher-level) strategies, which in turn provide stability, continuity, and direction for tactical behaviours. (Lewicki et al., 2004, pp. 110-111)

In our case we can infer that, in the beginning, cause–effect relationship would have been marked by *competition*. Both Begin and Sadat wanted peace in the Middle East, but their disposition to *collaboration* was weak. Their strategy was to try to get as much as possible from their opponent while yielding nothing or as little as possible of their own. Carter's strategy was to gently bring them to a position closer to *integration–collaboration*. He worked from what apparently was an *accommodation* plane, prioritising a good relational outcome over good substantive outcomes for securing the USA interests. With this strategy, he looked for points of deep connection, in order to build on them an agreement over the differences: "(...) The objective is to minimize their real differences and maximize their apparent similarities [underlined in the original text]" (Vance, 1978 –declassified August 13, 2003–, TAB 4 –Considerations for Conducting the Summit Meetings–, p. 1).

The mediator's strategy was necessary to gradually turn the symmetrical cause–effect relationship between Begin and Sadat into a less destructive, more constructive one. For a constructive relationship "(...) agreement is necessary on several issues: the price, the closing date of (...), renovations to—" (Lewicki et al., 2004, p. 38). Finally, reaching BATNA was, if *not* an absolute *collaboration* at least within a *bargaining mix*.

The process

The negotiation process can be summarized as the development of the 'transitive property', since the implementation of the strategy creates *not only* a 'symmetric' relationship, determined by the cause–effect relationship, but also a collateral effect, unaccounted for, produced by the tactical acts –a sort of third party.

Therefore, the process leading to accomplish some type of agreement, of a greater or lesser extent in accordance with the goals, depends largely on the process development and its effects on the participating actors.

Taking all that into account, together with the negotiation timing, we can now analyse each one of its phases[5]:

- Phase 1: *Preparation*. (February 14, 1977 to September 5, 1978). On February 14, Carter took the initiative of writing letters to Sadat and

---

[5] The *phases* are taken from Lewicki et al., 2004, p. 117, Figure 4.3.

Rabin (Israeli Prime Minister at that time), at the request of his State Secretary –Cyrus Vance– at his return from a mission in the Middle East, urging them to start peace negotiations for the area:

From the letter from President Jimmy Carter to President Anwar al-Sadat of Egypt, written during the first month of President Carter administration:

> (...) I look forward to meeting whit you personally at the earliest opportunity. I have asked Secretary Vance to discuss when and how our first meeting might be arranged. In view of the importance of Egypt in our common pursuit of peace (...). The growing friendship and cooperation between Egypt and the United States have already brought us some steps along the path to peace. (Camp David Accords: Twenty-Five Documents After Twenty-Five Years, 2003. Document 1, 1977, pp. 1-2 –declassified on October 3, 1997–)

From the letter from President Jimmy Carter to President Yitzhak Rabin of Israel, written during the first month of President Carter administration:

> I am confident that USA–Israeli relations will continue on the cordial and sound basis that has characterized the close ties between our countries and peoples for three decades. I look forward to working closely whit you in our common search for a lasting peace settlement in the Middle East. (Camp David Accords: Twenty-Five Documents After Twenty-Five Years, 2003. Document 2, 1977, p. 1 –declassified on December 11, 1998–)

From that day (February 14) to the start of the negotiations in Camp David (September 6, 1978) the countries involved prepared the negotiation meticulously, since the interests at stake were of paramount importance to each one of them, as we have already seen.

- Phase 2: *Relationship building*. The relationships among the negotiators were modelled during the *Preparation Phase*. We should remark here that the position of the USA was that of an impartial mediator. For this, Carter had a private meeting with each of the other two negotiators in the morning of September 6, with the aim to create an atmosphere of confidence during the negotiations:

    > (...) seeking to build a common recognition of the <u>unique opportunity</u> these talks offer, the <u>responsibility to history</u> the three of

> them share, and the need to grasp the nettle now [underlined in the original text]. He could emphasize that the negotiations have reached a stage where only heads of government can break the impasse, and therefore each side must try understand the other's political problems. (...) Since each side will have as an objective capturing the USA, the President with sympathy for each side's interests *will have to establish the independence of the USA position* [emphasis added]. Each will want a sense of special relationship with us; we will want to be close to each without being in either's pocket. (Vance, 1978 –declassified on August 13, 2003–, TAB 2 –A Scenario for Camp David–, pp. 2-3)

- Phase 3: *Information gathering*. This is the time when all the negotiators meet to explain to each other what has brought them to this gathering and what do they expect to get from it. This meeting took place in the afternoon of September 6:

> Later in the day, the three men used the patio outside Aspen for further discussions. They talked about three issues: 1) the Sinai Peninsula between Egypt and Israel, 2) the ownership of the West Bank and Gaza areas bordering Israel, and 3) the role that Palestinian people would have in governing themselves. (Camp David Accords: Thirteen Days After Twenty-Five Years, 2003, September 6, 1978 – gathered for the 25th anniversary of the Camp David Accords)

- Phase 4: *Information using*. Here the negotiators will express themselves with regard to the way in which an agreement will be possible on the three points expounded in the former phase. Along this phase the constraints/preferences for an eventual agreement will become apparent. This was the hardest part of the negotiation, because Sadat and Begin were directly affected by the subjects to be discussed: sovereignty, security, and economy (the three pillars of foreign policy), with the aggravating circumstance of having to talk about Jerusalem, Holy City for Jews and Muslims, and a necessary subject when discussing the territories occupied during the '67 War. This phase started in the morning of Thursday September 7 and was finished by Carter in the night of Tuesday September 12 when, in view of the sterility of endless discussions, he decided to assertively intervene by choosing the less problematic issue, the Sinai, to set a framework for an agreement: " "I decided to work that afternoon on the terms for an Egyptian–Israeli treaty, and spread the Sinai maps out on the dining table to begin this task, writing the proposed agreement on a yellow scratch pad." —Jimmy Carter from *Keeping Faith*" (Camp David Accords: Thirteen Days After Twenty-Five Years, 2003, September 12, 1978 – gathered for the 25th anniversary of the Camp David Accords).

- Phase 5: *Bidding*. This was the most difficult time of the negotiation. The positions of Sadat and Begin seemed irreconcilable and Carter feared *not* to reach an agreement. But peace in the Middle East was crucial for the USA interests and he decided to 'impose' peace in the region:

> Determined to reach agreement on a framework for peace, Carter and Vance spent eleven hours with Aharon Barak from Israel and Osama el-Baz from Egypt to work out the detailed language of the framework proposal. As they hammered out the language of each phrase, both Barak and el-Baz demonstrated their astute legal minds and their excellent knowledge of English. When differences in language stopped progress, President Carter suggested that "West Bank" be used in the English and Arabic texts, while "Judea and Samaria" be used in the Hebrew version; "Palestinians" in the English and Arabic, yet "Palestinian Arabs" in the Hebrew. He would explain the change in a letter to Begin. The letter would be attached to any formal agreement they would reach. The letter exchange idea became a critical factor in making progress toward agreement. (Camp David Accords: Thirteen Days After Twenty-Five Years, 2003, September 13, 1978 –gathered for the 25th anniversary of the Camp David Accords)

In spite of Carter's efforts, the negotiation is in a deadlock of *bargaining–bargaining*:

> "We can go no further." –Carter
> "I am leaving." –Sadat
> President Sadat could not agree to leave Israeli settlements and airfields in the Sinai Peninsula, and Prime Minister Begin could not agree to remove these settlements. Without agreement on these issues, there did not appear to be any way to continue. Carter had already told the delegations that Sunday, September 17, would be the last day of the meetings. He had requested that all the delegations work on a joint statement about the meetings, emphasizing the positive accomplishments. (Camp David Accords: Thirteen Days After Twenty-Five Years, 2003, September 15, 1978 –gathered for the 25th anniversary of the Camp David Accords)

In this critical situation, with only 48 hours left, Carter played his *only* winning trick –reminding Sadat of the importance of *collaborating* with Israel and changing his own attitude in order to reach a peace agreement, since future Egypt–USA relationships depend on this:

> "I explained to [Sadat] the extremely serious consequences… that his action would harm the relationship between Egypt and the United States, he would be violating his personal promise to me… [and] damage one of my most precious possessions –his friendship and our mutual trust." —Jimmy Carter from *Keeping Faith* [additions included in the original text]. (Camp David Accords: Thirteen Days After Twenty-Five Years, 2003, September 15, 1978 –gathered for the 25th anniversary of the Camp David Accords)

Next morning he addressed Begin's iron position:

> "Ultimatum, Excessive Demands, Suicide" –Begin
> Even though the progress of the talks was faltering, Carter's determination to reach agreement remained strong. In another negotiating session with Begin, Barak, and Dayan, Carter and Vance made a case for peace, going through the Sinai framework and the Framework for Peace line by line. (Camp David Accords: Thirteen Days After Twenty-Five Year, 2003, September 16, 1978 –gathered for the 25th anniversary of the Camp David Accords)

- Phase 6: *Closing the deal*. Carter's words to Sadat and Begin had the expected effect: on that same day –September 16– a safe path was tended toward the peace agreement:

> Carter explained to Begin that Sadat would not continue negotiations toward a peace treaty until the Israeli settlements in the Sinai region were removed. After a storm of protest, Begin finally agreed to submit the question of settlements to the Israeli Knesset for a decision—*If any agreement is reached on all other Sinai issues, will all the settlers be withdrawn?* [emphasis in the original text] He even promised to allow each Knesset and Cabinet member to vote individually, without the requirements of political party loyalty. This was acceptable to Sadat!
> Carter explained to Sadat that Begin would not allow the phrase "inadmissibility of acquisition of territory by war" to be part of the Framework for Peace. [1967 U.N. Resolution 242, which contains this phrase, is to be found in the annex of the Framework. Begin

claimed that it did not apply to Israel because the 1967 War was a defensive war for his country] [square brackets in the original text]. Begin insisted that only permanent residents of the West Bank and Gaza areas, not all Palestinians, participate in future peace negotiations. Sadat agreed to write one letter defining Egypt's role in these negotiations and one letter stating his position on an undivided Jerusalem. This was acceptable to Begin!

All through the meetings, Carter continued to remind Sadat and Begin how much each had to gain in making peace. (Camp David Accords: Thirteen Days After Twenty-Five Years, 2003, September 16, 1978 –gathered for the 25th anniversary of the Camp David Accords)

- Phase 7: *Implementing the agreement*: In the end, there was a 'formula', the 'integral' had been defined, a point of convergence had been reached: the incenter of our triangle. "… a significant achievement in the cause of peace…", in the Carter's words(Camp David Accords: Thirteen Days After Twenty-Five Years, 2003, September 17, 1978 – gathered for the 25th anniversary of the Camp David Accords, September). This was the moment to define how, when and where would the agreement be implemented. This was the moment to draw the circumference to circumscribe the encounter point –incenter. On Sunday September 17, 1978, two accords were reached: a *Framework for Peace in the Middle East* and a *Framework for the Conclusion of a Peace Treaty between Egypt and Israel*. Both were signed by the three leaders – Anwar al-Sadat, Menachem Begin and Jimmy Carter– on the same day, at the official signing ceremony of the *Camp David Accords* in Washington.

At this point of our analysis, the pieces of the puzzle are already in place. Sadat and Begin were able to move from a *distributive* position to an *integrative* one, thus reaching an agreement that met the goals. But there is still one more step to go in our analysis process because, as every integral, this one has also a differential coefficient; *What is the 'coefficient' for this 'integral'?* This is the question we will address next.

The Key Piece: The Agreement. Level of Adjustment

The 'differential coefficient' will allow us understanding the agreement of the negotiation. Our question now should be what has been the 'differential coefficient' here, and why this 'coefficient' –and *only* this one– has made it possible to reach the goal: an accord for peace in the Middle East. Ultimately, the 'differential' is the crux of the matter, since it is what gives us the key to give a logical answer to the question posed in the introduction: *Why were the agreements these, and **not** others?* Then, let us analyse the final documents of *Camp David Accords*.

From the Camp David Accords: The Framework for Peace in the Middle East (September 17, 1978):

Preamble
The search for peace in the Middle East must be guided by the following:
- *The agreed basis for a peaceful settlement of the conflict between Israel and its neighbors is United Nations Security Council Resolution 242, in all its parts* [emphasis added].
- After four wars during 30 years, despite intensive human efforts, the Middle East, which is the cradle of civilization and the birthplace of three great religions, does not enjoy the blessings of peace. The people of the Middle East yearn for peace so that the vast human and natural resources of the region can be turned to the pursuits of peace and so that this area can become a model for coexistence and cooperation among nations.
- The historic initiative of President Sadat in visiting Jerusalem and the reception accorded to him by the parliament, government and people of Israel, and the reciprocal visit of Prime Minister Begin to Ismailia, the peace proposals made by both leaders, as well as the warm reception of these missions by the peoples of both countries, have created an unprecedented opportunity for peace which must not be lost if this generation and future generations are to be spared the tragedies of war.
- The provisions of the Charter of the United Nations and the other accepted norms of international law and legitimacy now provide accepted standards for the conduct of relations among all states.
- To achieve a relationship of peace, in the spirit of Article 2 of the United Nations Charter, future negotiations between Israel and any neighbor prepared to negotiate peace and security with it are necessary for the purpose of carrying out all the provisions and principles of Resolutions 242 and 338.
- *Peace requires respect for the sovereignty, territorial integrity and political independence of **every state** in the area and their right to live in peace within secure and recognized boundaries free from threats or acts of force* [emphasis added]. Progress toward that goal can accelerate movement toward a new era of

reconciliation in the Middle East marked by cooperation in promoting economic development, in maintaining stability and in assuring security.
- Security is enhanced by a relationship of peace and by cooperation between nations which enjoy normal relations. In addition, under the terms of peace treaties, the parties can, on the basis of reciprocity, agree to special security arrangements such as demilitarized zones, limited armaments areas, early warning stations, the presence of international forces, liaison, agreed measures for monitoring and other arrangements that they agree are useful.

As we see here, the compromise to abide by UN SC RES 242 (1967) and UN SC RES 338 (1973) is to be reached *only* after previous settlement with each of the concerned States; a negotiation with every one of them was necessary. However, after this first document was issued, *only one* more document, focusing on Egypt, could be issued, since Egypt was the *only* represented party at the negotiations. That is how the second Camp David document came: *Framework for the Conclusion of a Peace Treaty between Egypt and Israel.* Let us find out what this document has to say about our initial question *Why were the agreements these, and **not** others?*

From the Camp David Accords: Framework for the Conclusion of a Peace Treaty between Egypt and Israel (September 17, 1978):

It is agreed that:
- The site of the negotiations will be under a United Nations flag at a location or locations to be mutually agreed.
- All of the principles of U.N. Resolution 242 will apply in this resolution of the dispute between Israel and Egypt.
- Unless otherwise mutually agreed, terms of the peace treaty will be implemented between two and three years after the peace treaty is signed.
- The following matters are agreed between the parties:
1. *the full exercise of Egyptian sovereignty up to the internationally recognized border between Egypt and mandated Palestine* [emphasis added];
2. *the withdrawal of Israeli armed forces from the Sinai* [emphasis added];
3. the use of airfields left by the Israelis near al-Arish, Rafah, Ras en-Naqb, and Sharm el-Sheikh for civilian purposes only, including possible commercial use only by all nations;
4. the right of free passage by ships of Israel through the Gulf of Suez and the Suez Canal on the basis of the Constantinople Convention of 1888 applying to all nations; the Strait of Tiran and Gulf of Aqaba are

international waterways to be open to all nations for unimpeded and nonsuspendable freedom of navigation and overflight;
5. the construction of a highway between the Sinai and Jordan near Eilat with guaranteed free and peaceful passage by Egypt and Jordan; and
6. the stationing of military forces listed below.

Now it is possible to understand why the presence of Egypt was necessary but *not* sufficient to create and reach a working agreement in favour of achieving peace. *Where is the key for sufficiency?* This key lays precisely in mentioning which were the territories belonging to Egypt before the *'67 War*: the Sinai and Gaza. It is now when we are ready to ask the big question: *Why did the Sinai come back to Egypt and **not** any other of the territories occupied during the '67 War?* The answer is just a few lines above: "1. the full exercise of Egyptian sovereignty up to the internationally recognized border between Egypt and mandated Palestine" (Camp David Accords: Framework for the Conclusion of a Peace Treaty between Egypt and Israel, September 17, 1978). At this point, it seems that everything is solved and the puzzle has been perfectly assembled; but incisive minds may have still another question: *Why does Egypt acknowledge the limits to its sovereignty at the border between itself and the Palestinian territories?* This is the precise meaning of Egypt acceding to recover the Sinai but *not* Gaza. The answer is to be found in *al Quran*, in the *Surah* which Begin mentioned to Sadat when he learned of his wish to visit Jerusalem (November 11, 1977):

> Your President said, two days ago, that he will be ready to come to Jerusalem, to our Parliament - the Knesset - in order to prevent one Egyptian soldier from being wounded. It is a good statement. I have already welcomed it, and it will be pleasure to welcome and receive your President with the traditional hospitality you and we have inherited from our common father, Abraham. And I, for my part, will, of course, be ready to come to your capital, Cairo, for the same purpose: *No more wars - peace - a real peace, and for ever. It is in the Holy Koran, in Surah 5, that our right to this Land was stated and sanctified. May I read to you this eternal Surah* [emphasis added]:

"Recall when Moses said to his people: Oh my people, remember the goodness of Allah toward you when He appointed prophets amongst you.... Oh my people, enter the Holy Land which Allah hath written down as yours ... [emphasis added]"[6]
It is in this spirit of our common belief in God, in Divine Providence, in right and in justice, in all great human values which were handed down to you by the Prophet Mohammed and by our Prophets –Moses, Yeshayahu, Yermiyahu, Yehezkiel- it is in this human spirit that I say to you with all my heart: Shalom. It means Sulh. (Camp David Accords: Twenty-Five Documents After Twenty-Five Years, 2003. Document 5, 1977, p. 2)

Now, we ought to round-up the former question, in an almost rhetorical manner: *Why did Begin remain this particular Surah to Sadat? What was the pre-existing foundation that validates his words?* The answers is to be found in the Holy Torah: *"In that day the* LORD *made a covenant with Abram, saying: 'Unto thy seed have I given this land, from the river of Egypt unto the great river, the river Euphrates* [emphasis added]" (Genesis. 15:18).[7] Given that, the *only one* territory occupied by Israel during the *'67 War* beyond the biblical bounders was the Sinai, so it was the *only one* that could come back in strict observance of Surah 5:22-28 and Genesis 15:18.[8]

---

[6] For wider information, here is the transcription of the complete text from *al Quran* that Begin quoted. Due to its key role in the perception of our case, we have preferred to take the text in Spanish, an accurate translation from the original text, rather than risking a free translation to English or taking an unwarranted on line English version of the original Arabic. *Surah 5, 22-28*: "22. Cuando Moisés dijo a los israelitas: acordaos de los beneficios que habéis recibido de Dios; ha suscitado profetas en vuestro seno, os ha dado reyes, os ha concedido favores que no ha concedido jamás a nación ninguna. 23. Entra, ¡oh pueblo mío!, en la tierra santa que Dios te ha destinado; no volváis atrás por temor a que os encaminéis a vuestra perdición. 24. Este país, respondieron los israelitas, está habitado por hombres poderosos. Mientras lo ocupen, nosotros no entraremos en él. Si salen, nosotros tomaremos posesión de él. 25. Presentaos a la puerta de la villa, dijeron los hombres que temían al Señor y que estaban favorecidos por sus gracias: no bien hayáis entrado, seréis vencedores. Poned vuestra confianza en Dios, si sois fieles. 26. ¡Oh Moisés!, dijo el pueblo, no entraremos mientras no haya salido el pueblo que la habita. Ve con tu Dios y combatid ambos. Nosotros permaneceremos aquí. 27. Señor, exclamó Moisés, solo tengo poder sobre mí y sobre mi hermano; pronuncia sobre nosotros y este pueblo de impíos. 28. Entonces el Señor dijo: Esta tierra les estará prohibida durante cuarenta años. Andarán errantes por el desierto, y tú cesa de atormentarte a causa de este pueblo de impíos" (Al-Kharat, 2007).
[7] The Jewish Publication Society of America (1917).
[8] Argument on the importance of the Biblical allotment of land to this negotiation has been verified in a personal communication –private meeting– with a high representative of the Jewish religious community in Belgium who preferred to remain anonymous, December 13, 2010.

As it has been shown, the level of adjustment –the 'differential' in our 'integral'– has been reduced to an explanation/quantification as a one-variable function, *culture*, and specifically *religion*, the deepest and most radical

---

[9] Retrieved from http://www.bible.ca/archeology/bible-archeology-exodus-kadesh-barnea-southern-border-judah-territory-river-of-egypt-wadi-el-arish-tharu-rhinocolu.htm, checked on October 11, 2019.

component of culture. *So is it that, in the beginning of the XXI century, religion still has something to teach us?* The answer may *not* be to believe in God, its object, but to believe in religion in itself, its subject, for there are still today human beings and peoples that live in strict observance of religious laws.

CONCLUSION: Logic in the Past, a Lesson for the Future

Once our case, *Camp David Accords*, has been fully analysed, only one question remains: *Is the effort of this minute analysis a contribution, or is it just an intellectual divertimento?* To answer this question we have to go back to the reflection on history that was done in the introduction, where the argument was that the only possibility to go on building History is through a profound understanding of the reasons leading to a certain agreement, and *not* to a different one, in the course of an international negotiation. After all, History is just the history of the disagreements and agreements attained by mankind throughout the ages.

Thus, the humble contribution of this investigation is to highlight the role of culture from the angle of each negotiator's beliefs, *not only* religious, but of the type "I believe X, and *not* Y". Culture, the natural channel to reveal and to transmit beliefs, shapes us in an unyielding manner. Let us look again at the definition of *culture* in the New Encyclopædia Britannica (1990):

> *Culture, the integrate pattern of human knowledge, belief and behaviour* [emphasis added]. Culture thus defined consists of language, ideas, beliefs, customs, taboos, codes, institutions, tools, techniques, works of art, rituals, ceremonies and other related components; *and the development of culture depends upon man's capacity to learn and to transmit knowledge to succeeding generations* [emphasis added]. (Vol. 3, p. 784. Quoted by Lang, 1993, p. 38)

We can now say that the culture (a part of which are the beliefs of the actors present in an international negotiation) is the cornerstone to prepare a negotiation and resolve it reaching a possible and plausible agreement. This is to recognize that negotiation is a science in itself, as it requires to accurately study all the pieces involved, but it is also an art. Negotiations demand a *savoir faire* that is *not* related to deduction from tangible knowledge, but to abduction from inductive knowledge including science and art. As Lewicki et al. (2004) say:

> *The notion that negotiation is both art and science is especially valid at the cross-cultural or international level* [emphasis added]. The science of negotiation provides research evidence to support broad trends that often, but not always, oc-

cur during the negotiation. The art of negotiation is deciding which strategy to apply when, and choosing which models and perspectives to apply to increase cross-cultural understanding. *This is especially challenging because cross-cultural and international negotiations add a level of complexity significantly greater than within-culture negotiations* [emphasis added]. (p. 443)

A deep knowledge of the cultures present in a negotiation is a great vantage point to reach a final agreement, because its success is *not* to obtain what was initially desired, but to obtain the best result that the pieces in this particular game can provide. Aiming at what these pieces can provide, and *not* at particular desires, is more realistic and scientific, and also less frustrating. It is necessary *not* to dismiss a single one of the variables involved in the game, including the least visible but more present one: culture, the niche where beliefs dwell; this is the decisive variable, and, as sugar in the coffee, you do not see it but it is there.

Anyway, the main goal of this work is *not* whether it reached the desired purpose, but whether the reflections presented open the way to future thinking, the gates to future History. The ultimate purpose of this paper is to widen the horizon to the scrutiny of those minds which are passionate about negotiation. Our last question, dedicated to those who read this far, is:

what do you *think, believe*...?

Bibliography

MANUALS:

Moiffat AL-KHARAT, *El Corán*. Arganda del Rey (Madrid), Editorial EDIMAT Libros, S. A., 2007.

Michael BRECHER, *Decisions in crisis. Israel, 1967 and 1973*. California, California University Press, 1980.

Michael BRECHER, *A Study of Crisis*. Michigan, The University of Michigan Press, 1997.

Yoram DINSTEIN, *The International Law of Belligerent Occupation*. Cambridge, Cambridge University Press, 2009.

J. R. GAINSBOROUGH, *The Arab–Israeli Conflict*. Great Britain, Richard Clay Ltd., 1986.

Roy LEWICKI, Bruce BARRY & David SAUNDERS, *Essentials of Negotiation*. New York, McGraw-Hill, 2004.

John Norton MOORE, *The Arab–Israeli Conflict. Regarding and Documents*. Princeton (New Jersey), Princeton, 1977.

THE JEWISH PUBLICATION SOCIETY OF AMERICA, *The Holy Scriptures according to the Masoretic Text*. A new translation with the aid of previous versions and with constant consultation of Jewish authorities. Chicago, The Lakeside Press, 1917 (5677). Retrieved from https://jps.org/wp-content/uploads/2015/10/Tanakh1917.pdf, checked on October 11, 2019.

UNITED STATES CONFERENCE OF CATHOLIC BISHOPS, *The New American Bible*. Translated by Members of the Catholic Biblical Association of America. Encino (California), Benziger Editor, 1970.

William ZARTMAN & Maureen R. BERMAN, *The Practical Negotiator*. New Haven and London, Yale University Press, 1982.

BOOKS:

Charles ENDERLIN, *Paix ou Guerres. Les secrets des négociations israélo–arabes 1917-1995*. Paris, Fayard, 2004.

Etian HABER, Zeev SCHIFF & Ehud YAARI, *L'Année de la Colombe. Jérusalem 1977 — Camp David 1978*. Paris, Hachette, 1979.

Yitzakh RABIN, *Mémoires*. Paris, Éd. Buchet/Chastel, 1980.

Pierre RAZOUX, *La Guerre des Six Jours (5-10 juin 1967). Du mythe à la réalité*. Paris, Éd. Economica, 2006.

Samuel SEGUEV, *La Guerre de Six Jours. Opération « Drap Rouge »*. Paris, Calmann-Lévy, 1967.

BOOK CHAPTERS and ARTICLES:

John G. CROSS, *"Economic Perspective"*. In: *International Negotiation. Analysis, Approaches, Issues,* by Victor A. KREMENYUK (ed.). San Francisco, Jossey-Bass Inc. Publishers, 1991, pp. 164-179.

Chistophe DUPONT & Guy-Olivier FAURE, *"The Negotiation Process"*. In: *International Negotiation. Analysis, Approaches, Issues,* by Victor A. KREMENYUK (ed.). San Francisco, Jossey-Bass Inc. Publishers, 1991, pp. 40-57.

Oliver FAURE & Jeffrey Z. RUBIN, *"Lessons for Theory and Research"*. In: *Culture and Negotiation: The Resolution of Water Disputes,* by Guy Oliver FAURE & Jeffrey Z. RUBIN (eds.). Newbury Park (California), SAGE Publications, Inc., 1993, pp. 209-231.

Jean F. FREYMOND, *"Historical Approach"*. In: *International Negotiation. Analysis, Approaches, Issues,* by Victor A. KREMENYUK (ed.). San Francisco, Jossey-Bass Inc. Publishers, 1991, pp. 121-134.

Victor A. KREMENYUK, *"A Pluralistic Viewpoint"*. In: *Culture and Negotiation: The Resolution of Water Disputes,* by Guy Oliver FAURE & Jeffrey Z. RUBIN (eds.). Newbury Park (California), SAGE Publications, Inc., 1993, pp. 47-54.

Winfried LANG, *"A Professional's View"*. In: *Culture and Negotiation: The Resolution of Water Disputes,* by Guy Oliver FAURE & Jeffrey Z. RUBIN (eds.). Newbury Park (California), SAGE Publications, Inc., 1993, pp. 38-46.

Dayle E. POWELL, *"Legal Perspective"*. In: *International Negotiation. Analysis, Approaches, Issues,* by Victor A. KREMENYUK (ed.). San Francisco, Jossey-Bass Inc. Publishers, 1991, pp. 135-147.

Dean G. PRUITT, *"Strategy in Negotiation"*. In: *International Negotiation. Analysis, Approaches, Issues, by* Victor A. KREMENYUK (ed.). San Francisco, Jossey-Bass Inc. Publishers, 1991, pp. 78-89.

Jeffrey Z. RUBIN, *"The Actors in Negotiation"*. In: *International Negotiation. Analysis, Approaches Issues,* by Victor A. KREMENYUK (ed.). San Francisco, Jossey-Bass Inc. Publishers, 1991, pp. 90-99.

Jeswald W. SALACUSE, *"Implications for Practitioners"*. In: *Culture and Negotiation: The Resolution of Water Disputes,* by Guy Oliver FAURE & Jeffrey Z. RUBIN (eds.). Newbury Park (California), SAGE Publications, Inc., 1993, pp. 199-208.

James K. SEBENIUS, *"Negotiation Analysis"*. In: *International Negotiation. Analysis, Approaches, Issues,* by Victor A. KREMENYUK (ed.). San Francisco, Jossey-Bass Inc. Publishers, 1991, pp. 203-215.

Victor M. SERGEEV, *"Metaphors for Understanding International Negotiation"*. In: *International Negotiation. Analysis, Approaches, Issues,* by Victor A. KREMENYUK (ed.). San Francisco, Jossey-Bass Inc. Publishers, 1991, pp. 58-64.

I. William ZARTMAN, *"Regional Conflict Resolution"*. In: *International Negotiation. Analysis, Approaches, Issues,* by Victor A. KREMENYUK (ed.). San Francisco, Jossey-Bass Inc. Publishers, 1991a, pp. 302-314.

I. William ZARTMAN, *"The Structure of Negotiation"*. In: *International Negotiation. Analysis, Approaches Issues,* by Victor A. KREMENYUK (ed.). San Francisco, Jossey-Bass Inc. Publishers, 1991b, pp. 65-77.

I. William ZARTMAN, *"A Skeptic's View"*. In: *Culture and Negotiation: The Resolution of Water Disputes,* by Guy Oliver FAURE & Jeffrey Z. RUBIN (eds.). Newbury Park (California), SAGE Publications, Inc., 1993, pp. 17-21.

I. William ZARTMAN, *"Conflict Resolution and Negotiation"*. In: *The SAGE Handbook of Conflict Resolution,* by Jacob BERCOWITCH, Victor KREMENYUK & I. William ZARTMAN (eds.). London, SAGE Publications, Ltd., 2009, pp. 322-339.

DOCUMENTS:

Camp David Accords: Framework for the Conclusion of a Peace Treaty between Egypt and Israel (September 17, 1978). Retrieved from
https://www.jimmycarterlibrary.gov/research/framework_for_the_conclusion_of_a_peace_treaty

Camp David Accords: Jimmy Carter Reflects 25 Years Later (September 16, 2003). Retrieved from
http://www.cartercenter.org/news/documents/doc1482.html

Camp David Accords: Related Correspondence [annex to the framework agreements]. Retrieved from
https://www.jimmycarterlibrary.gov/research/camp_david_accords_related_correspondence

Camp David Accords: The Framework for Peace in the Middle East (September 17, 1978). Retrieved from
https://www.jimmycarterlibrary.gov/research/framework_for_peace_in_the_middle_east

Camp David Accords: Thirteen Days After Twenty-Five Years (September 2003. Gathered to commemorate the twenty-fifth anniversary of the Camp David Accords). Retrieved from
https://www.jimmycarterlibrary.gov/research/thirteen_days_after_twenty_five_years

Camp David Accords: Twenty-Five Documents After Twenty-Five Years (September 2003. Compiled to commemorate the twenty-fifth anniversary of the Camp David Accords). Retrieved from
https://www.jimmycarterlibrary.gov/research/twenty_five_documents_after_twenty_five_years

Document 1 and *Document 2*: These letters from President Jimmy Carter to President Anwar Sadat of Egypt (February 14, 1977) and Prime Minister Yitzhak Rabin of Israel (February 14, 1977) were written during the first month of President Carter administration. They indicate President Carter's early personal commitment to the Middle East peace process as well as his eagerness to meet with both Sadat and Rabin. *(1) Declassified on October 3, 1997. (2) Declassified on December 11, 1998.*

Document 3: This letter (June 28, 1977) from nine United States Senators represented the support that President Carter was to have in his quest for Middle East peace.

Document 4: There were to be many obstacles to Middle East peace, and President Carter appealed to President Sadat for his support in this letter (October 21, 1977). One month later President Sadat visited Israel for the first time. *Declassified on July 14, 1997.*

Document 5: President Sadat's plan to visit Israel solicited this speech by Israeli Prime Minister Menachem Begin (November 11, 1977), who had succeeded Prime Minister Rabin after a surprise election victory.

Document 6: This President Sadat letter to President Carter (April 24, 1978) recognized the international respect that was accorded to President Carter's success in negotiating the Panama Canal treaties. President Sadat also emphasized Israeli activities and negotiating positions that discouraged support of other Arab nations for the peace process. *Declassified on June 19, 2002.*

Document 7: One week later, as President Carter prepared to meet with Prime Minister Begin, National Security Adviser Zbigniew Brzezinski asserted (May 1, 1978) that a renewed commitment to Israeli security should be coupled with an appeal for Begin recognition of President Sadat's requirements. *Declassified on October 20, 1997.*

Document 8 and *Document 9*: On August 3, 1978, President Carter wrote private letters to President Sadat and Prime Minister Begin to be delivered by Secretary of State Cyrus Vance. The letters proposed a meeting of Carter, Sadat, and Begin at a time and location to be determined. *Both declassified on July 14, 1997.*

Document 10: President Carter's schedule for September 5, 1978, records the arrival at Camp David of First Lady Rosalynn Carter, President Sadat, and Prime Minister Begin.

Document 11: Ten days later (September 15, 1978) President Carter sent this message to Prime Minister Begin and President Sadat to suggest an honorable and respectable closure to what then appeared to be a failed peace effort.

Document 12: This is President Carter's draft of what became the "Framework for the Conclusion of a Peace Treaty between Egypt and Israel."

Document 13: President Carter's notes indicate the difficulty of the last few hours of negotiations at Camp David.

Document 14: President Carter's schedule for September 17, 1978, records the hectic activities that culminated in a signing ceremony in the East Room of the White House late on a Sunday evening.

Document 15: The next day (September 18, 1978) President Carter accepted the congratulations of former National Security Adviser and Secretary of State Henry Kissinger.

Document 16: On Sunday, September 17, 1978, cellist Mstislav Rostropovich had been hosted by First Lady Rosalynn Carter at a White House concert as Carter, Begin, and Sadat wrapped up their Camp David discussions.

Document 17: One month later Prime Minister Begin and President Sadat were awarded the Nobel Peace Prize. Both President Sadat and National Security Adviser Brzezinski believed that President Carter should have been included in the award, an honor that President Carter was eventually accorded in December, 2002.

Document 18: As this memo (November 21, 1978) from Brzezinski to the President suggests, tough negotiations continued after the Camp David Accords until March, 1979. *Declassified on October 20, 1997.*

Document 19: On March 7, 1979, the President and Mrs. Carter departed for Egypt and Israel to once again bring the President's personal force to bear on the post-Camp David negotiations of an Egyptian–Israeli peace treaty.

Document 20: This is the President's schedule on March 13, 1979, the day on which he finally nailed down a peace treaty agreement.

Document 21: The President and First Lady returned to Andrews Air Force Base shortly after midnight on March 14, 1979, to be welcomed back to the United States by an enthusiastic crowd of approximately one thousand people.

Document 22: March 26, 1979, featured the signing ceremony for the Treaty of Peace between the Arab Republic of Egypt and the State of Israel in the afternoon on the North Lawn of the White House, followed that evening by a State Dinner on the South Lawn of the White House.

Document 23: President Carter's handwritten editing of his statement delivered at the signing ceremony demonstrates his attention to detail even to the last moment of the peace process.

Document 24 (August 20, 1980) and *Document 25* (September 3, 1980): Despite the success of the Egyptian–Israeli Peace Treaty in preserving peace between those two nations, the Middle East continued to be a volatile region throughout the Carter Administration and to the present.

Key Legislature During the Carter Administration. Retrieved from https://www.jimmycarterlibrary.gov/research/keylegis

National Security Council (NSC) Staff and Organisation, 1977-81 [Carter Administration]. Retrieved from https://www.jimmycarterlibrary.gov/research/nsc_staff_and_organization_1977_81

The Egyptian–Israeli Treaty: text and selected documents. In: *Basic Documentary Series*, No. 13. Beirut, Institute for Palestine Studies, 1979.

UN SC RES 242 (November 22, 1967) Retrieved from https://undocs.org/S/RES/242(1967)

UN SC RES 338 (October 22, 1973) Retrieved from https://undocs.org/S/RES/338(1973)

Cyrus R. VANCE (1978). *Briefings to prepare the Camp David Accords* [declassified on August 13, 2003]. Retrieved from https://www.cartercenter.org/documents/nondatabase/campdavidstudy.pdf

INTERVIEWS:

David L. AARON (Deputy—National Security Council), *Exit Interview* (December 15, 1980) [declassified on June 15, 1994]. Retrieved from https://www.jimmycarterlibrary.gov/assets/documents/oral_histories/exit_interviews/Aaron.pdf

Zbigniew BRZEZINSKI (National Security Adviser), *Exit Interview* (February 20, 1981). Retrieved from https://www.jimmycarterlibrary.gov/assets/documents/oral_histories/exit_interviews/Brzezinski.pdf

Robert J. LIPSHUTZ (Counsel to the President), *Exit Interview* (September 29, 1979). Retrieved from https://www.jimmycarterlibrary.gov/assets/documents/oral_histories/exit_interviews/Lipshutz.pdf

WEBSITES:

https://www.cartercenter.org/ Authorized website of Jimmy Carter Center.

www.cidcm.umd.edu Authorized website of the Centre for International Development & Conflict Management (CIDCM) at the University of Maryland.

https://www.jimmycarterlibrary.gov/ Authorized website of Jimmy Carter's Presidency.

https://www.un.org/securitycouncil/ Authorized website of the United Nations Security Council.

CARTOGRAPHY:

http://www.bible.ca/archeology/ Website on biblical archaeology and geography.

Frédéric ENCEL, *Atlas Géopolitique d'Israël. Aspects d'une démocratie en guerre*. Paris, Éd. Autrement, 2008.

Author

M. Dolors Martínez-Cazalla (Barcelona, 1971) holds a PhD in Philosophy, Logic and Philosophy of Science (2017, University of Seville), a Master's Degree in Diplomacy and Conflict Resolution (2011, University of Louvain-la-Neuve), a Master's Degree in Protocol (2014, UNED –Open University of Spain), she is Neuro-Linguistic Practitioner student since 2015 (Centre of Excellence, Manchester), and has a long professional experience in trading negotiations.

She wrote her PhD, entitled *Negotiating with Logical-Linguistic Protocols in a Dialogical Framework*, under the supervision of Prof Dr Ángel Nepomuceno at the University of Seville and Prof Dr Shahid Rahman at the University of Lille 3. This book is based on her PhD and the discussions during the viva.

Martínez-Cazalla is currently associate member of the research group *Lógica, Lenguaje e Información* (GILLIUS –HUM609, University of Seville) and an active external collaborator in the *Axe Transversal Argumentation* headed by Shahid Rahman (laboratory *Savoirs, Textes, Langage* –Unité Mixte de Recherche 8163, Lille 3).

Main Publications:
[2018]: Hunting Hackers. A gift from the 'mute guest' (A study about how to pass information as safely –prudently– as possible). *Revista de Humanidades de Valparaiso*, 6 (12), 115-140.
doi: https://doi.org/10.22370/rhv.2018.12.1344;
[2017]: El protocolo la 'red invisible' en las relaciones internacionales y la diplomacia (Protocol, the strongest diplomacy weapon. A key piece in the international relations). *Revista Estudios Institucionales*, 4 (6), 99-116.
doi: https://doi.org/10.5944/eeii.vol.4.n.6.2017.18749;
[2016]: La 'fuga' de Don Proti (cuento) o La ubicación de los laicos de puesto nombrado en los actos oficiales de la Iglesia Católica (Tradition converted into legal canonical present. A challenge resolved). *Revista Estudios Institucionales*, 3 (4), 267-304.
doi: https://doi.org/10.5944/eeii.vol.3.n.4.2016.18385;
[2012]: La Importancia de las Creencias en las Negociaciones Internacionales (An exhaustive analysis about how the beliefs of the negotiators played a crucial role in the *Camp David Accords* –September 1978). In C. Morano-Rodríguez, J. Campos-Acosta, & M. M. Alcubilla-Martín (Eds.), *Ciencia, Humanismo y Creencia en una Sociedad Plural* (pp. 407-416). Oviedo, España: Ediciones de la Universidad de Oviedo y Fundación Castroverde.

madomartinezcazalla@gmail.com

www.ingramcontent.com/pod-product-compliance
Lightning Source LLC
Chambersburg PA
CBHW071624170426
43195CB00038B/2107